W9-CCA-426

Planet Earth

UNDERGROUND WORLDS

Other Publications:

COLLECTOR'S LIBRARY OF THE CIVIL WAR
LIBRARY OF HEALTH
CLASSICS OF THE OLD WEST
THE EPIC OF FLIGHT
THE GOOD COOK
THE SEAFARERS
THE ENCYCLOPEDIA OF COLLECTIBLES
THE GREAT CITIES
WORLD WAR II
HOME REPAIR AND IMPROVEMENT
THE WORLD'S WILD PLACES
THE TIME-LIFE LIBRARY OF BOATING
HUMAN BEHAVIOR
THE ART OF SEWING
THE OLD WEST
THE EMERGENCE OF MAN
THE AMERICAN WILDERNESS
THE TIME-LIFE ENCYCLOPEDIA OF GARDENING
LIFE LIBRARY OF PHOTOGRAPHY
THIS FABULOUS CENTURY
FOODS OF THE WORLD
TIME-LIFE LIBRARY OF AMERICA
TIME-LIFE LIBRARY OF ART
GREAT AGES OF MAN
LIFE SCIENCE LIBRARY
THE LIFE HISTORY OF THE UNITED STATES
TIME READING PROGRAM
LIFE NATURE LIBRARY
LIFE WORLD LIBRARY

FAMILY LIBRARY:
HOW THINGS WORK IN YOUR HOME
THE TIME-LIFE BOOK OF THE FAMILY CAR
THE TIME-LIFE FAMILY LEGAL GUIDE
THE TIME-LIFE BOOK OF FAMILY FINANCE

This volume is one of a series that examines the workings of the planet earth, from the geological wonders of its continents to the marvels of its atmosphere and its ocean depths.

Cover
A multiple-exposure photograph details the progress of a caver descending the Incredible Pit, a 440-foot natural shaft in Ellison's Cave in Georgia. Dangling precariously from a nylon rope and drenched by an underground waterfall, the caver triggered a hand-held flash four times to record the pit's awesome depth.

Planet Earth

UNDERGROUND WORLDS

By Donald Dale Jackson
and The Editors of Time-Life Books

Time-Life Books, Alexandria, Virginia

Time-Life Books Inc.
is a wholly owned subsidiary of

TIME INCORPORATED

FOUNDER: Henry R. Luce 1898-1967

For information about any Time-Life book, please write:
Reader Information
Time-Life Books
541 North Fairbanks Court
Chicago, Illinois 60611

First printing. Printed in U.S.A.
Published simultaneously in Canada.
School and library distribution by Silver Burdett Company, Morristown, New Jersey.

Library of Congress Cataloguing in Publication Data
Jackson, Donald Dale, 1935-
Underground worlds.
(Planet earth; 6)
Bibliography: p.
Includes index.
1. Caving—History—19th century. 2. Caving—History—20th century. 3. Caves. I. Time-Life Books. II. Title. III. Series.
GV200.62.J33 796.5'25 82-782
ISBN 0-8094-4320-1 AACR2
ISBN 0-8094-4321-X (lib. bdg.)
ISBN 0-8094-4322-8 (mail order ed.)

THE AUTHOR

Donald Dale Jackson is a former staff writer for *Life* and the author of four previous Time-Life books in the American Wilderness and Flight series. Among his other publications are a volume on the judicial system of the United States and a history of the California gold rush.

THE CONSULTANTS

Dr. John Holsinger is Professor of Biological Sciences at Old Dominion University in Norfolk, Virginia, and a Fellow of the National Speleological Society. A recognized specialist on cave invertebrates, he has published numerous articles on cave fauna.

Ernst Kastning, Assistant Professor of Geology at the University of Connecticut, has written extensively on the history of cave exploration and cave science. He is a Fellow of the National Speleological Society and has served as Director of Publications for the Cave Research Foundation.

Arthur N. Palmer is Professor of Geology and Director of the Water Resources Program at the State University of New York. He has spent more than a quarter of a century exploring, mapping and interpreting the geology of caves and karst regions throughout the European and North American continents.

CONTENTS

THE LURE OF THE ABYSS

"My hair stood on end, my teeth chattered, my limbs trembled," declared one of literature's most famous cavers as he entered a shaft and began the epic adventure recounted in Jules Verne's *Journey to the Center of the Earth.* Although Verne's tale is entirely fictional, he could not have captured better the misgivings many cavers feel as they step into dark and mysterious subterranean worlds. Yet apprehension is mixed with heady anticipation: Caves offer their visitors spectacular scenery, scientific marvels and, above all, the rare exhilaration of discovery.

Caves are the result of minute incremental etchings of a variety of natural processes that, over a period of many millennia, have hollowed cavities extending for scores of miles into the earth's interior. Passages wider than an interstate highway may feed into fissures that are breathlessly thin. Vertical shafts, often obscured by pitch-black shadows, plummet hundreds of feet. Yet here also are found some of nature's most exquisite artifacts. Beneath massive layers of limestone uplifted by some titanic convulsion, explorers may find a tiny shimmering crystal more fragile than spun glass.

By scrutinizing rock formations in caves, scientists have found clues to geologic events that occurred a billion years ago. And by studying the remarkable creatures that thrive in a cold, wet world of impenetrable darkness, these explorer-scientists have illuminated the wondrous adaptability of living things.

But as much as caves have given to science, they remain chiefly the domain of amateur explorers. Every year, thousands of cavers probe the underground frontier, seeking passages and chambers such as the ones shown here and on the following pages. Guided by small lights and a keen sense of adventure, they thread their way through perilous mazes with a persistence that confirms Jules Verne's observation: "There is nothing more powerful than this attraction toward an abyss."

A team of cavers exploring the nine-mile-long labyrinth of St.-Marcel d'Ardèche in southeastern France pause to survey the grandeur of a central gallery. The cave, carved out by the Ardèche River system tens of thousands of years ago, still offers tantalizing opportunities for discovering unknown passages.

With two waterproof flashlights strapped to his helmet, a diver emerges from the water that almost fills a gallery in the Cave of Caumont in northern France. Diving equipment is so cumbersome that cavers take it along only when they are certain to need it.

A caver wades through a shallow underground stream in an illuminated passage in Virginia's Nutt Cave. As the stream gently but persistently erodes the floor of the cave, water seeping through the ceiling slowly deposits bristling rows of stalactites.

Struggling through the aptly named Agony Crawlway in Ellison's Cave in Georgia, an explorer hauls thick coils of strong rope to be used for a pit descent. Not far beyond The Agony lies The Fantastic Pit—at 510 feet, the longest vertical drop in U.S. caves.

Suspended by one rope and steadied by a second, a caver lowers himself into a deep pit in San Agustín Cave in Mexico. Explorers routinely negotiate dizzying vertical shafts, using mountain-climbing techniques.

In Phanonga, Thailand, a series of limestone terraces totaling 20 to 30 feet in height supports a delicate shroud of flowstone. The formation is the result of mineral-laden water spilling over the terrace edges, depositing microscopic particles with each tiny surge.

A 20-foot column surrounded by rapier-like stalactites connects the floor and ceiling of the Dome Room in New Mexico's Carlsbad Cavern. Such columns form when stalactites above meet stalagmites on the floor—a process that may take 100,000 years.

Deep within Mammoth Cave in Kentucky, a caver pauses to examine slabs of rock that have fallen from the cave ceiling. Because it signals the end of the cave's life cycle, this collapse of rock strata is called cave breakdown.

Chapter 1

EXPLORERS OF A STYGIAN REALM

Andy Eavis, a 30-year-old professional engineer employed by a Yorkshire colliery, discovered the first clue to the existence of a major cave in Borneo in late 1977. At the time, he was sitting comfortably at home in Selby, England, perusing aerial photographs of a rain forest on the other side of the world. Eavis was also the chairman of the International Speleological Union's exploration section, and the Royal Geographical Society had invited him and five other cave experts to join an expedition into the newly designated Gunong Mulu National Park in the Malaysian state of Sarawak on northern Borneo. While more than 100 scientists—including botanists, zoologists, ecologists, geologists, anthropologists—were to study the pristine rain forests aboveground and prepare a management plan for the park, Eavis and his comrades would probe its subterranean honeycomb of caves.

The photographs of Mulu's unmapped jungle, which was populated only by nomadic Penan tribesmen, were dominated by a jagged, three-mile-wide limestone ridge that extended some 20 miles in a northeast-southwest direction in north Sarawak. The ridge, whose peaks reached as high as 6,000 feet, was a perfect example of what geologists call karst, the terrain formed by deeply eroded limestone. Most of the erosion is done by water; rather than shedding rain water, karst absorbs it like a sponge. For two million years heavy tropical rains had been sculpting the limestone into a strange, craggy landscape where rivers suddenly vanished into the ground, yawning sinkholes opened in the earth and rock bridges spanned whole canyons.

It was obvious to any experienced caver that this terrain concealed a labyrinth belowground. After surveying a few easily accessible Sarawak caves for the Malaysian government in 1961, geologist Gerald E. Wilford had reported that "large spectacular caves are most likely to be discovered in the uninhabited and relatively unexplored Melinau area between the Tutoh and Limbang Rivers"—a region within the 210-square-mile park.

Eavis and his colleagues called themselves simply cavers (cave cognoscenti detest the term "spelunker," although scientifically inclined cavers do style themselves speleologists), but they were scarcely amateurs; the Royal Geographical Society had vetted both the caving experience and scientific credentials of these six before inviting them to join the expedition. Team members boasted professional expertise in geology, geomorphology, cave biology and hydrology. Among them, they had mapped and studied every major cavern in Europe, descending shafts hundreds of feet deep and spending days underground. Their skill in rock-climbing techniques—belaying, prusiking, rappelling and the like—surpassed that of most mountaineers.

By the eerie glow of torches, members of an 18th Century expedition inspect a cave near Cornial, in what is now Yugoslavia. The group, led by Austrian explorer Joseph Nagel, had been ordered to map the limestone caves of the area by Holy Roman Emperor Francis I.

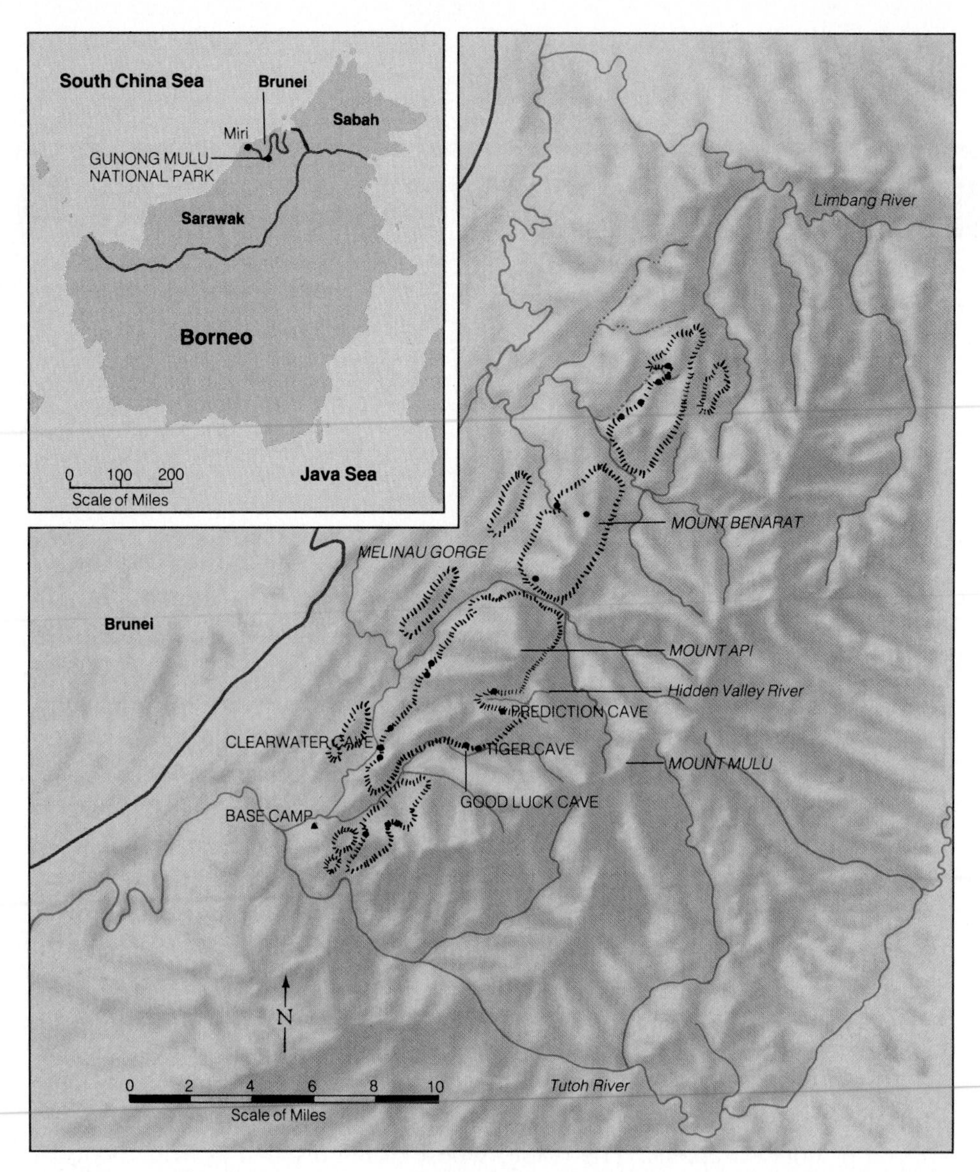

Gunong Mulu National Park, site of extensive British caving expeditions in 1978 and 1980, is located in Sarawak in northwestern Borneo *(inset)*. The explorers concentrated their efforts on a 20-mile-long, three-mile-wide limestone ridge system *(hatched outline)* that contains more than a dozen caves *(dots)*.

As he studied stereoscopic photographs through a special viewer, Eavis spotted any number of cave entrances in the Melinau Limestone, the geological formation that made up part of the ridge that interested him. He was particularly intrigued by a huge limestone overhang on one of the ridge's peaks above a river that seemed to abruptly sink into the ground. Because the overhang was above a deep valley, Eavis knew that a prehistoric river many times the size of the present stream had once flowed through the valley and under the cliff, dissolving the limestone beneath the overhang—and perhaps leaving an enormous cave along its subterranean course.

The hypothesis remained in the back of his mind during the ensuing months, as the cavers attended to the details involved in mounting any wilderness expedition: immunizations, solicitations of food and equipment, travel arrangements, fund raising. They arrived at their Mulu National Park base, about 60 miles southeast of Miri on the southern end of the limestone ridge, in March 1978. Local porters and tribesmen had assured expedition members that there were no caves in the region, but the cavers decided to pursue Eavis' hunch even though it would be an arduous gamble.

Eavis and two colleagues were guided to the site of the putative cave by the leader of the scientific expedition, Robin Hanbury-Tenison, who had already scouted the area. With four porters, they slogged northeastward for two days through the knee-deep muck of a steamy flood plain, quickly becoming accustomed to the jungle's hazards: a debilitating combination of

85° F. heat and 100 per cent humidity; constant assaults by buzzing clouds of mosquitoes and sand flies; encounters with leeches, centipedes, scorpions and king cobras; and not least of all the ravages of *borak,* a potent homemade rice wine forced on the cavers by friendly tribesmen.

Toward the end of their 10-mile trek they forded raging rivers in the foothills and finally climbed the knife-edged ridge, 3,000 feet high at this point, on the southern end of Gunong Api, or Mount Api. They then descended into the valley in which the overhang was located, dubbed it Hidden Valley, and set up camp. Because their group was too small to mount a rescue operation should disaster strike, they set up short-wave radio communications with the base camp. Next day, Eavis and the other two cavers scrambled several hundred feet up the precipitous valley wall through rocky debris and thick vegetation to the massive overhang. Under it they found a steep slope, slick with mud and water, plunging 600 feet into the mountain. There the entrance opened into a soaring overhead shaft 200 feet in diameter and perhaps 1,000 feet high. The team named their discovery Lubang Ramalan—Prediction Cave—in honor of Eavis' shrewd prophecy.

Exploring the cave beyond the shaft was a sweaty, risky ordeal. The jagged limestone could slice like a razor through cloth and flesh, and the standard safety procedure was to wear protective coveralls. Struggling along in the tropical heat was, according to one caver, "like having an airless, hot shower." Torrential rains frequently generated subterranean flash floods that could trap or drown cavers. Even dried bat guano on the cave floor could be dangerous: It harbored fungus spores that cause histoplasmosis, a potentially deadly fever.

To all these hazards Prediction Cave added special obstacles. The passage narrowed to a crawlway that was 300 feet wide but only two feet high, its bottom littered with pebbles and wet clay. It was, one caver wrote later, "one of the most ridiculous passages under the earth." For a seemingly interminable time, the cavers slithered on their bellies through this black crack, pulling their gear behind them in canvas bags, guided only by their flickering carbide helmet lamps. After struggling thus for 400 feet they reached a crumbling chamber nearly filled by an 80-foot mound of boulders. The cave seemed to continue into the mountain on the other side of the mound—a trench extended in that direction—but the cavers were at a dead end. Another pile of fallen boulders, a choke, filled the passageway.

The explorers did not yet despair, but instead resorted to geological detective work to reconstruct the cave's history. A huge river apparently had been flowing through the cavern when the roof collapsed and restricted its flow with the boulder choke. The slower-moving water deposited sediment more rapidly, eventually filling the giant tunnel to its roof. During eons of time the river carved its way deeper into Hidden Valley, the water receded from the cave and the sediment gradually settled, creating a cave floor that bore an imprint of the contours of the roof. By this reasoning it seemed likely that a passage—perhaps an enormous one—lay beyond the choke.

The ancient river must have flowed out of the mountain somewhere, and the cavers scoured its flanks for days but did not find the missing link to Prediction Cave. Instead, a few miles to the southwest, they discovered two other caverns, Tiger Cave and Clearwater Cave, both with impressive streams flowing from their mouths. If these streams were resurgences of the river in Hidden Valley, the existence of a large cave network in the ridge—although not necessarily the one they had entered at Prediction Cave—would be confirmed. To find out, the cavers watched the water emerging from Clearwater Cave while porters dumped 44 pounds of a fluorescent green dye into the river near Prediction Cave. The observers at Clearwater

Mulu: The Genesis of a Cave Region

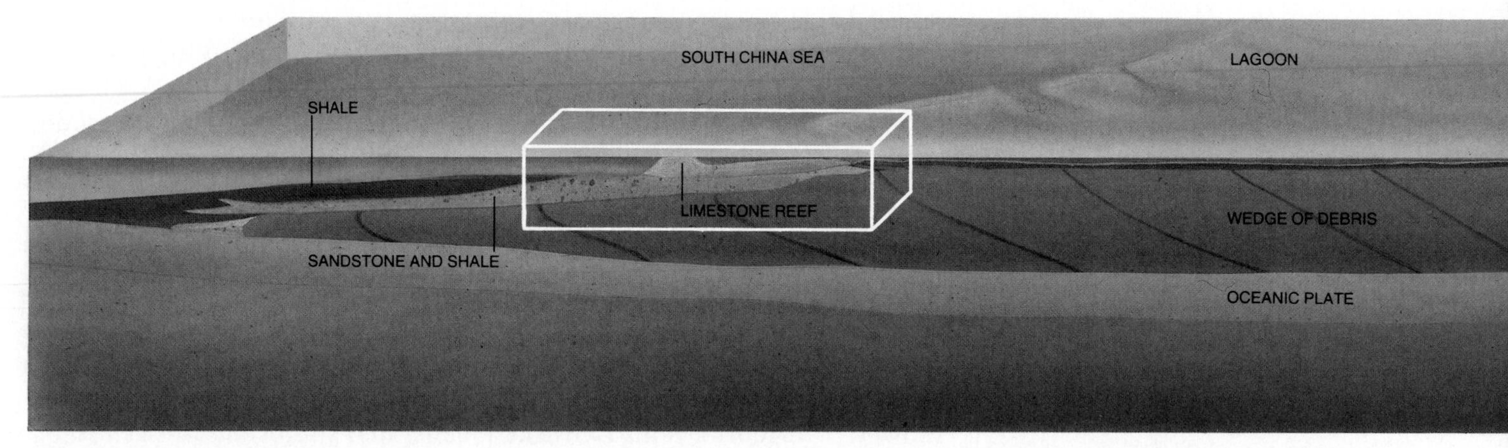

The magnificent caves of the Mulu region in northwest Borneo exist because of a remarkable sequence of gentle sedimentation and titanic collisions that began some 60 million years ago. At that time, one of the earth's great slabs of crustal material, or tectonic plates, began to dive under another plate carrying the land mass now known as Borneo.

As the plate ground downward, thick layers of sediment were scraped off and piled up at the edge of the land mass. This wedge of debris accumulated slowly beneath the sea and year after year was blanketed by the remains of billions of marine animals. This ages-long submarine blizzard of calcite sea shells created thick drifts of new sediment, some of which were eventually compressed into a limestone barrier reef on top of the wedge of debris *(above)*. Concurrently, mud and clay on the surrounding deep-ocean floor consolidated into shale.

Millions of years later, further movements of the plates caused this entire area to sink, and the limestone was buried by layers of shale and other sediments *(right, top and middle)*. About five million years ago, however, the tectonic dynamics changed; the diving plate rose and began to grind more directly against the overriding plate. The awesome energies of this collision buckled and lifted the rock layers at the edges of the plates until a new shale-capped mountain was thrust above the sea.

After two million years, heavy tropical rains had worn off the shale, the highly soluble limestone was uncovered and erosion accelerated both on the surface and deep within the rock. During the past million years, erosion whittled the mountain into three separate peaks *(right, bottom)* that dominate the Mulu landscape today: Mount Mulu, made of sandstone and shale, and the twin limestone pinnacles of Api and Benarat, which house the spectacular Mulu caves.

An artist's conception of the conditions 40 million years ago shows in cross section an area of the northwest coast of present-day Borneo. The oceanic plate carrying the South China Sea floor is being subducted beneath Borneo, and layers of sedimentary rock, including a limestone reef, are forming offshore.

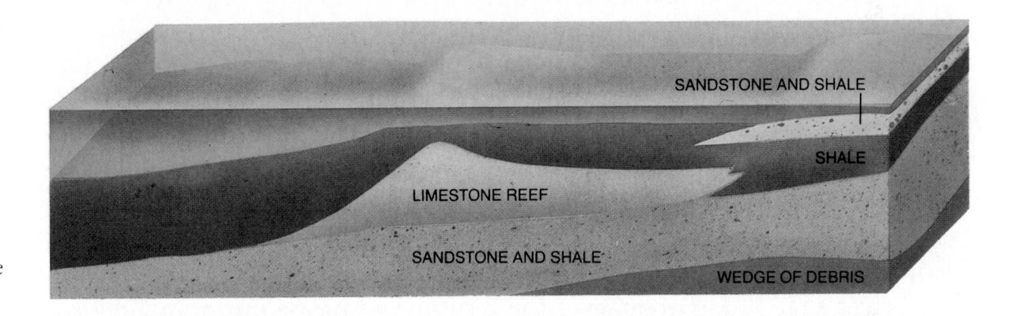

A close-up of the area where the Mulu caves would emerge portrays the geological situation about 20 million years ago. The limestone reef has subsided beneath the sea and is buried under shale. Sediments of coarser-grained sandstone and shale have begun to wash into the area from inland Borneo to the southeast.

The stresses exerted by the moving plates, accompanied by earthquakes, cause irregular folding of the layers of shale and sandstone. By 10 million years ago, the younger sediments from Borneo blanketed the entire reef complex.

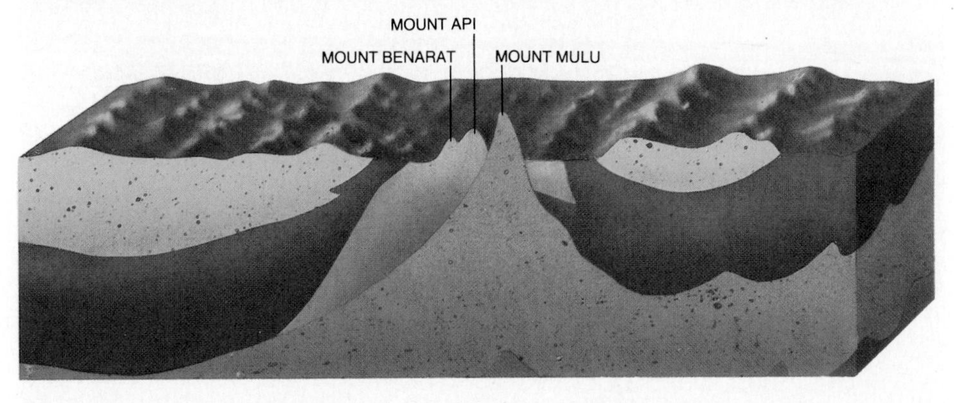

Continued uplifting raises the reef complex high above sea level. It took about two million years for the shale cap of the mountain to erode; in the ensuing one million years, the limestone eroded so fast that the pinnacles of Api and Benarat, where the caves are located, are now about 2,500 feet lower than the 8,000-foot sandstone-and-shale peak of Mount Mulu.

thought they saw traces of dye floating past hours later, but the test was inconclusive. The dye apparently had been diluted to invisibility in the network of underground rivers. Still, the circumstantial evidence favoring a connection convinced most of the cavers, and when they later prepared a map of the 30 miles of caves they had explored, they included an underground drainage from the Hidden Valley river to Clearwater Cave.

The matter might have ended there, but Eavis still believed there was a larger, undiscovered cavern beyond Prediction Cave. In December 1980, he and Ben Lyon, warden of a British caving center, led a 22-man expedition back to Borneo. This expedition was supported by the Sarawak Forestry Department, which wanted to establish Mulu as a caving area and train local people in caving techniques. Besides exploration and scientific study, the objectives included the production of a documentary film on the caves.

Eavis promptly organized a search for the elusive cave. Last-minute logistical problems—a crate of essential gear had failed to arrive in Borneo—prevented him from conducting the search. But on December 28, 1980, Hans Friederich, a Dutch hydrologist, began a reconnaissance on his behalf. After resurveying Tiger Cave, Friederich began methodically searching the tangled cliff line above the river flowing from the cave mouth. He stumbled on a hidden resurgence barely 400 yards away—a slit in the face of the cliff, perhaps 15 feet wide and 150 feet high, with a water-filled canal at its base. A tremendous wind blowing from the passage suggested that it might lead to a major cave, but Friederich did not have the time or equipment to explore farther, so he trekked the 10 miles back to camp for reinforcements.

The next morning Eavis, Friederich and a group of porters commenced a scientific adventure that quickly took on aspects of a comic opera. Heavily laden with caving hardware—lights, 300-foot ropes, surveying tapes and instruments—they hiked back to the passage. Only then did Eavis learn that the entrance was full of deep water, a detail Friederich had forgotten to mention. Without life jackets or inflatable dinghies, the cavers could not even enter the cave. But they did conduct a new, and this time definitive, dye test. Four and a half hours after porters dumped red dye into the river near Prediction Cave, the resurgence turned crimson, proving that the unexplored passage was the primary conduit from Hidden Valley. The cavers named the passage Lubang Nasib Bagus, Malay for "Good

Local porters, one in an inflatable dinghy and the other knee-deep in water, navigate the 330-yard-long entrance canal in Borneo's Good Luck Cave. The men were part of the expedition that discovered the cave in December 1980.

A caver hauls himself up a slippery bank above a 50-foot-wide plunge pool at the head of a series of rapids in Good Luck Cave. Members of the 1980 expedition had to swim across the pool before they could begin the final approach to Sarawak Chamber.

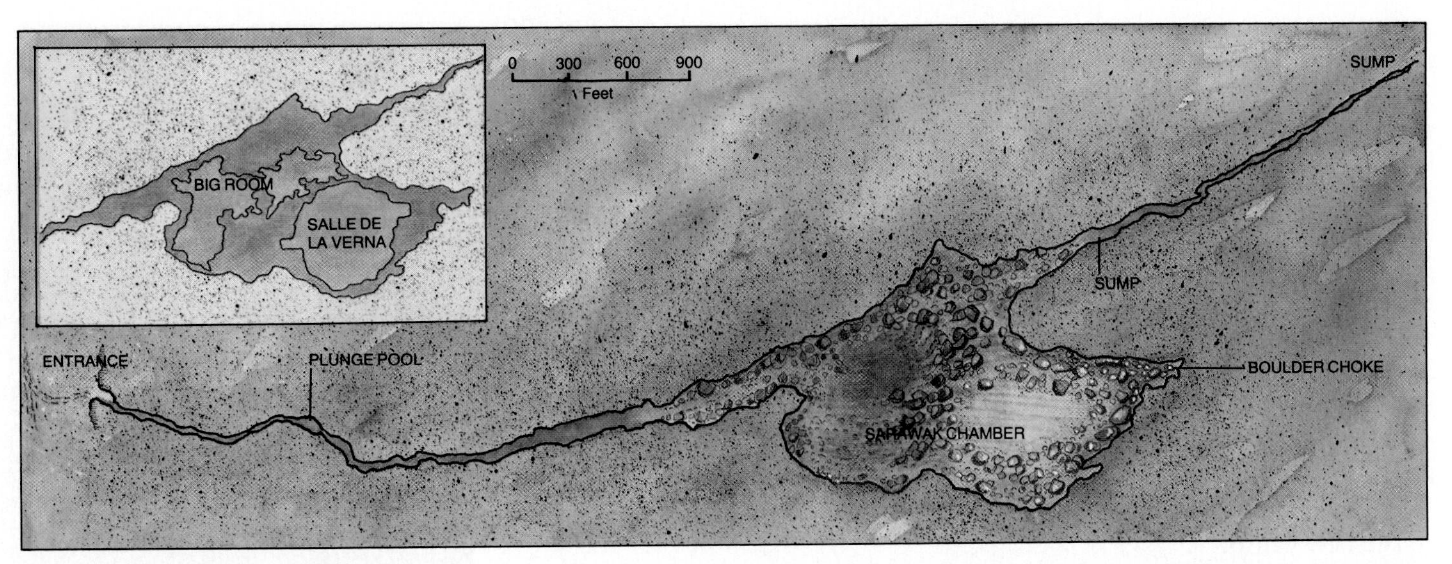

Good Luck Cave extends some 8,000 feet beneath the limestone mountains of Mulu National Park in Borneo. Its Sarawak Chamber is so enormous that the world's second- and third-largest underground chambers—Carlsbad Cavern's Big Room and Pierre St.-Martin's Salle de la Verna—both could fit in it with room to spare *(inset)*.

Luck Cave," to commemorate Friederich's luck in finding the missing link.

After an overnight bivouac the cavers returned to camp for the proper gear—and were confined there by heavy rains. In three days 24 inches fell, raising river levels by 10 feet, submerging the floor of the base long house under three inches of water. For several days the cavers had to use boats simply to move between buildings; travel farther afield was impossible.

As the flood began to recede, the team mounted an exploration in the grand style, to survey what they expected to be a major discovery. Two cavers, the three-man film crew and six porters crowded onto a Malay longboat, a huge log dugout with a protective awning amidships and an outboard motor for propulsion. The overloaded craft traveled up the swollen streams as far as it could, until the propeller began fouling on trees and stumps, then dropped the cavers and their equipment on a jungled shoreline and headed back to camp. As the outboard's growl faded into the distance, the cavers began following the shore toward the resurgence—and found themselves walking in circles. They had landed on an island. The disgruntled explorers gingerly forded the waist-deep torrent, then set up a spartan temporary camp in a nearby cave. Renewed rains promptly marooned them there for two days.

When they finally overcame their misfortunes and reached Good Luck Cave on January 6, 1981, the team set off up the canal in two inflatable dinghies. Eavis and Friederich paddled slowly upstream ahead of the film crew. As the little flotilla voyaged along the arrow-straight canal, surveying the passage with tape and compass as they went, the cave roof receded into darkness overhead. The easy boat trip ended 1,500 feet inside the cave, at the foot of a series of rapids descending from the darkness.

With the film crew trailing behind, Eavis and Friederich cautiously edged upward through the slippery rapids for 250 feet to a swirling emerald green pool at their head, which they named the Plunge Pool. Here the film crew recorded scenes of the cavers swimming through the turbulent water and climbing the steep face at the pool's inlet, a tedious process that entailed considerable posing and reshooting. The entire company then returned to the surface and bivouacked at the cave entrance.

After a few hours of sleep Eavis and Friederich reentered the cave alone to explore beyond the Plunge Pool, where the cave abruptly narrowed to a sheer, towering canyon with the river coursing a tortured path along its narrow floor. The only route through this black chasm was a slow, risky, horizontal traverse along the canyon wall. A slip could send a caver hurtling onto the rocks 100 feet below. By the light of their carbide helmet lamps, the pair strung safety guidelines across tricky sections until, perhaps 400 feet into the traverse, they had paid out their entire supply of rope. They faced a smooth, blank wall that demanded safety lines; reluctantly, they returned to the cave entrance.

The next day Eavis and Friederich carried large coils of extra rope to the canyon, clipped themselves to the safety lines with metal carabiners and rapidly traversed the explored section, then laboriously clambered spider-like across the sheer face, hammering metal bolts into the rock to secure the safety line. Just beyond, the canyon widened, and they returned to the river, which broadened and soon disappeared: It was emerging from the cave floor at the base of a steep uphill slope. As the pair began climbing the slope, another caver caught up with them, shouting that a deluge outside was beginning to flood the canal. The three men hastily backtracked yet again to the surface. Although the torrent quickly subsided, days of delay had exhausted the team's food supply, and they were forced to return to the base camp.

On the following day Eavis broadcast an enthusiastic description of the new cave over the short-wave radio network that connected the expedition's far-flung outposts. He estimated that the cave was 200 feet wide—not the largest passage in Mulu, but quite impressive. Within 24 hours the base was swarming with speleologists from other camps, all clamoring to explore Good Luck Cave. The dinghies could hold only three cavers with their bulky gear, so the assembled company drew straws for the right to accompany Eavis. The honor was won by Dave Checkley, a veteran cave biologist, and Tony White, regarded as one of the world's premier cave explorers.

The three cavers quickly retraced the route through the Plunge Pool and the 650-foot traverse, then began climbing the steep, steadily widening slope, which was covered with sandy sediment, scree and boulders. As they ascended, they hastily surveyed the cave, since the expedition rarely had the time or manpower for a separate survey after exploration was complete. From each starting point one caver clambered uphill to the end of a 30-meter steel tape while another measured direction with a compass and the elevation angle with a clinometer; the third took notes and paced off the approximate dimensions of side passages and other landmarks.

At the top of the 1,300-foot slope, about three quarters of a mile from the cave entrance, the voices of the cavers began to echo—and the echoes continued for about 10 seconds before the absolute stillness returned. Unwittingly, they had stepped from the passageway into a large chamber, and had lost contact with its walls. They decided to follow a straight compass course until they reencountered a wall. When they found one, 320 feet due south, it was made of shale. That meant they were now under the Melinau Limestone and had found geological proof that the cave was at the bottom of the limestone layer, on the impermeable bed of the prehistoric river.

Surveying the undulating wall proved exceedingly difficult. The steep uphill climb was littered with huge limestone blocks bigger than houses, some so precariously balanced that they shifted when touched. "We often started surveying round a boulder, thinking it was the passage wall, only to find that the real wall was one hundred feet behind it," recalled Eavis. "Then we'd have to go back and resurvey."

After 12 hours and 77 survey legs of 30 meters each, the trio reached a branch passage barely 100 yards long. At its end they found a boulder choke and felt a strong draft blowing out from it—a sure sign of a connection to the surface. The trench on the north side of the passage and the size and shape of the boulders matched the features Eavis had seen two years before in Prediction Cave. Here, surely, was the missing link to Hidden Valley.

Barely discernible against the glare of the powerful lights they are holding, members of the British exploration team manage to illuminate only a portion of the huge Sarawak Chamber. One team member later described his awe at "hearing the incredible echo and feeling the immensity in every direction."

Expedition members clamber around a boulder to measure a segment of Sarawak Chamber. Surveying the 2,300-by-1,300-foot cave floor, with its profusion of house-sized boulders, took 16 hours of uninterrupted work.

A 17th Century engraving based on a painting by Flemish artist David Teniers depicts Satan tempting Saint Anthony to abandon his life of asceticism in a cave. Teniers's perception of a cave as a terrifying nether world filled with macabre creatures was typical of his time.

To this point the cavers, groping blindly through the dark void, had assumed that they were in a large passage that wound through the mountain, with an unseen wall roughly parallel to the one they had been following. Tony White now suggested that they had instead been surveying a huge chamber, but the others doubted the notion. "Dave and I thought it just wasn't possible to get a chamber this size," Eavis said. "We knew from compass bearings that the wall was bending round but we assumed there was another wall maybe 300 feet away."

After several more hours and 23 laborious survey legs, the trio finally reached a narrow passage where the upper end of the underground river appeared, after meandering beneath the cave floor for a half mile. They followed the river until it flowed into a sump—an impassable reach where the cave roof touches water—more than two miles from the cave entrance. The day's 9,000-foot survey had exhausted the three cavers. Both Eavis and Checkley were suffering from an excruciating tropical fungus infection that turned their feet a flaming red—the cavers called it "Mulu Foot"—and wanted only to get out quickly. But White, still convinced that they were in a gargantuan chamber, proposed returning to the entrance along a straight compass bearing to the southwest, a route that inevitably would run into the opposite wall of a curving passage.

So the trio marched out into the dark expanse, maintaining a compass course through a maze of blocks and boulders until they reached a level, sandy plain, the signature of an underground chamber. The sudden awareness of the immensity of the black void caused one of the cavers to suffer an acute attack of agoraphobia, the fear of open spaces. None of the three would later reveal who panicked, since silence on such matters is an unwritten law among cavers. "I have been in all of the world's big chambers," Eavis said later, "but this was different. One of us definitely did not like it and would not go far from a wall. There is a real fear of getting lost when you are wandering among boulders 300 feet across." After the victim recovered, the cavers began an easy stroll across the open, sedimented area; it was now absolutely clear that they had been in a single enormous void throughout their surveying.

The men emerged after 16 hours underground, torn between exhaustion and exultation, to announce that they had found the world's largest subterranean chamber—three times the size of the previous record holder, New Mexico's Carlsbad Cavern. A subsequent survey confirmed that Sarawak Chamber, as it was named in honor of the Malaysian state of Sarawak, was 2,300 feet long, 1,300 feet wide and never less than 230 feet high—big enough to contain 17 football fields. "Its dimensions can be gained from the survey, but not the spirit of the place," one explorer wrote afterward. "This was a strange world that no writer of fiction would invent."

Yet there is no assurance that even Sarawak Chamber is the world's largest. Unlike glaciers, mountains, oceans and other surface features, caves are secret places that cannot be found by satellite scans or airborne instruments. Many are geological infants that have no connection to the surface and can be discovered only by drilling or seismic sounding. The entrances to mature, open caves must be located by arduous foot searches through wild karst country. And the search is far from complete: Of the 5 to 10 per cent of the earth's land surface that is covered by karst, perhaps two thirds has been superficially examined for caves.

The earth's underground voids offer one of the last frontiers for old-fashioned explorers who crave adventure in the black unknown. And their very inaccessibility renders caves invaluable to scientists, both as museums and as laboratories. Within them lie records of precipitation, surface temperature and mineral formation writ over tens of thousands of years. From cave fossils, paleontologists learn about prehistoric animals such as the cave bear, woolly mammoth and bison; by studying the adaptation and mutation of the creatures that live in caves today, biologists test theories of evolution. Archeologists study the tools, sculpture and magical wall paintings of cave-dwelling primitive man.

Amaterasu, the shy Japanese sun-goddess, is enticed from her hiding place in a cave by a host of lesser divinities. According to legend, Amaterasu had retired to the cave in protest after her brother Susanowo, the storm-god, desecrated her rice fields.

Recognition of the scientific value of caves did not dawn until relatively recently. For most of history, the subterranean realm was associated less with knowledge than with mystery and marvels. According to mythology, a few gods, saints and heroes resided there: Zeus was born in a cave; the Japanese sun-goddess Amaterasu hid in a cave, plunging the world into darkness; King Arthur, his knights and his hounds are said to slumber in a Welsh cave, awaiting a visitor who will summon them again to battle. But mythological caves were largely peopled by mischief-makers and demons—trolls, gnomes, hobgoblins and elves.

Even the sophisticated ancients who struggled to replace superstition with rational thought believed, as the poet Vergil put it, that "there is something supernatural here." The Roman philosopher Seneca could not suppress his revulsion: He reported that a party of Greek silver prospectors who ventured underground "saw huge rushing rivers, and vast still lakes, spectacles fit to make them shake with horror. The land hung above their heads and the winds whistled hollowly in the shadows. In the depths, the frightful rivers led nowhere into the perpetual and alien night." After the miners' return to the surface, wrote Seneca, "they live in fear, for having tempted the fires of Hell." A 17th Century writer planning a visit to a Somerset cavern named Wookey Hole worried that the experience might permanently alter his disposition. "Although we entered in frolicksome and merry," he mused, "yet we might return out of it Sad and Pensive, and never more be seen to Laugh whilst we lived in the world."

But if these apprehensive visitors exaggerated the terrors of caves, it was a pardonable excess. The gloom, the strange apparitions that loomed at the edge of the glow cast by torch or candle, the unfamiliar and alarming sounds caused by the slap and spill of distant water echoing down rocky corridors, the deep pits that lay in wait for the unwary—this was a world of intimidating obstacles that most people were content to avoid.

The few who refused to be daunted usually had purposes other than adventure. Entrepreneurs probed several German caves during the 16th and 17th Centuries in search of unicorn horns, which were coveted for their purported medicinal value. Fortune hunters who thought they had succeeded in their quest unwittingly hauled to the surface the fossil remains of hyenas, bears and other animals that had died in their cave lairs. The effect on the patients to whom the resulting potions were given is not recorded.

There were more thoughtful explorations of caves during the 17th Century: for instance, those of John Beaumont, a Somerset surgeon and an amateur student of mining and geology. When in 1674 lead miners excavating a shaft in the Mendip Hills accidentally pierced a dome-shaped chamber, as sometimes happened in limestone country, Beaumont hastened to the site and hired six miners to accompany him into the cave. Carrying candles, the company descended the 60-foot shaft to the first chamber, which Beaumont proceeded to measure: It was 240 feet long, 8 feet wide and 30 feet high. "The floor of it is full of loose rocks," wrote Beaumont in his subsequent report to the Royal Society, England's most prestigious scientific organization. "Its roof is firmly vaulted with limestone rocks, having flowers of all colours hanging from them which present a most beautiful object to the eye"—apparently the same glittering stalactites and curtains of flowstone that awe visitors today.

The intrepid surgeon then led a 300-foot crawl through another narrow, rock-strewn passage, halting at the brink of a cavern so vast, Beaumont reported, that "by the light of our candles we could not fully discern the roof, floor, nor sides of it." Although the miners were familiar with the underground, they adamantly refused to descend this chasm, even for double pay. So Beaumont went down himself: "I fastened a cord about me, and ordered them to let me down gently. But being down about two fathom I found the rocks to bear away, so that I could touch nothing to guide myself by, and the rope began to turn round very fast, whereupon I ordered the miners to let me down as quick as they could." He landed dizzy but safe 70 feet below, on the floor of a cavern 115 feet in diameter and nearly 120 feet high, where he found large veins of lead ore.

Beaumont also visited nearby caves and returned several times to Lamb Leer, as the first cave was known, to supervise the digging of a horizontal shaft that revealed another large chamber. But his published accounts failed to stimulate much curiosity about caves. After a brief flurry of lead mining, Lamb Leer was abandoned; its entrance shaft eventually collapsed, leaving no trace, and the cave's very existence was virtually forgotten until the chamber was discovered anew in 1880.

Although investigations similar to Beaumont's took place elsewhere in Britain and on the Continent, exploring caves remained an eccentric, hit-or-miss pursuit. The early explorers rarely published reports on their work, so subsequent investigators could not build on old discoveries and techniques. The sole exception to this spasmodic pattern emerged in a region of Slovenia called by its Austrian overlords the Karst. (The name, referring to a 100-mile-wide limestone belt running along the coast of the Adriatic Sea, became the generic term for cave country in the 19th Century.)

One of the earliest Slovenian explorers was Baron Johann Valvasor, a well-traveled nobleman and amateur scientist who lived about 50 miles northeast of Trieste. During the 1670s and 1680s, Valvasor visited 70 caves

Between 10,000 and 40,000 years ago when the Great Ice Age was finally relaxing its grip on the earth, a new breed of human emerged who enjoyed all of the mental and physical attributes of modern man, including an unprecedented artistic ability. Known as Cro-Magnons after the site in France where their remains were first discovered in 1868, these beings left as their principal legacy a variety of vibrant paintings that have been preserved through the millennia on cave walls and ceilings.

Both moisture and the expansion and contraction caused by changes in the temperature cause pigment to deteriorate; thus the best-preserved paintings have been found in the driest and most remote regions of caves, where the temperature is constant. Why prehistoric hunters, who did not live in the depths of caves, ventured there to create thei art is a mystery. Some archeologists believe that the paintings were a form o hunting magic; others think they were ancestral totems or fertility symbols The answer may never be known.

Sadly, the very discovery of these prehistoric treasures has hastened their destruction. Microorganisms introducec from the outside world encourage the development of algae and bacteria on the cave murals. Already, such pollution has dulled many ancient paintings and, ir some cases, inflicted a greater amount o damage in just 10 years than the cave's natural environment had caused in thousands of years.

The profile of a bison, outlined by a cave artist 14,000 years ago in Niaux Cave in Ariège, France, still evokes the bulk and power of the ancient animal. Centuries of flowstone accumulation have partially obscured the sketch.

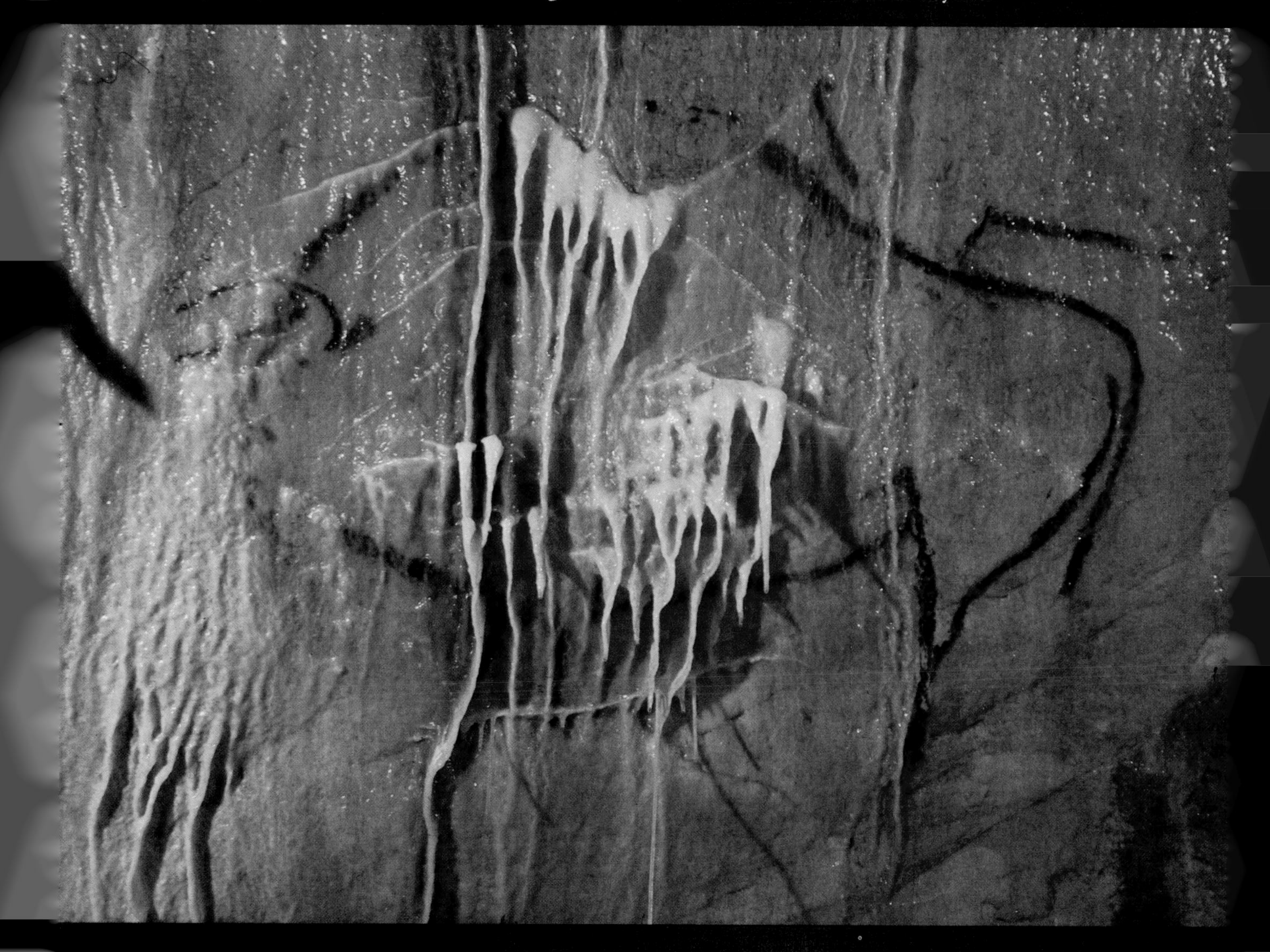

Adorned with silhouettes of the artist's hands, a prehistoric frieze discovered in Pech-Merle Cave in southern France features a pair of dappled horses standing back to back.

Sticklike human figures appear to be herding guanacos, wild relatives of llamas, in this prehistoric Indian painting found in a cave in Santa Cruz, Argentina. The handprints—a motif seen frequently in cave art—were made by placing the hand against the wall and using a tube to blow pigment over it.

An enigmatic pictograph, including a series of concentric circles, adorns a cave near Bustamante, Mexico. Early Indian tribes left such markings throughout North America.

A large wild cow leaps toward a lattice-like design, possibly representing a trap, in this mural in Lascaux Cave in southern France. Beneath the cow, three small horses gallop away.

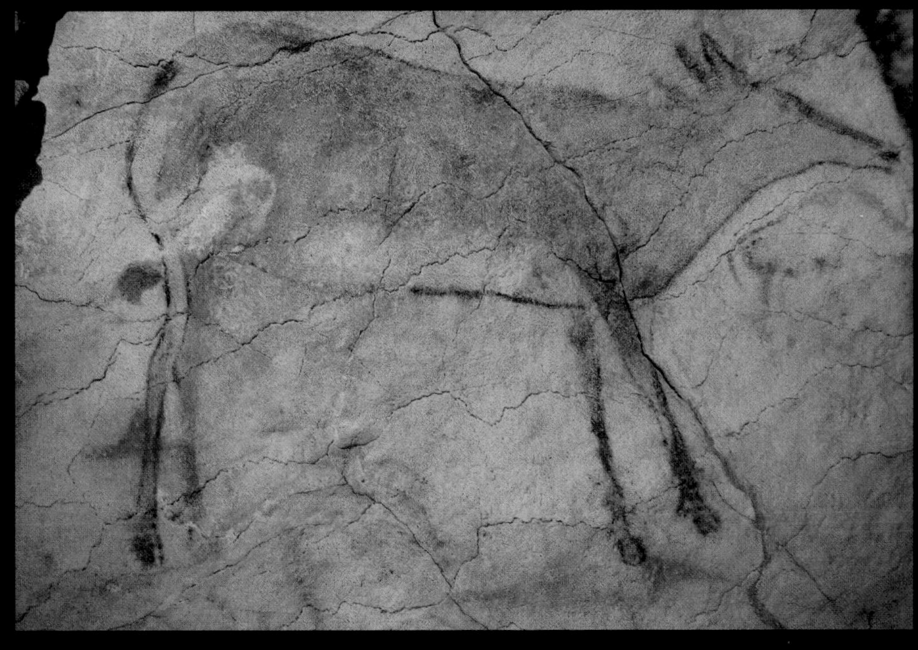

This seven-foot-long image of a female red deer (a small bison can be distinguished beneath its throat) occupies a corner of the ceiling in Altamira Cave in northern Spain.

in the Karst. He wrote meticulous reports on his discoveries, illustrated with sketches and maps, and eventually published them in a four-volume, 2,800-page collection. The cave dimensions in this document are unreliable. Valvasor estimated distances without measuring, and he tended to exaggerate—in one case, maintaining that he had explored nearly six miles of a cave that actually was less than half a mile long. But his work was remarkably comprehensive: He was the first to undertake a systematic series of cave explorations and to investigate the underground flow of water within and between caves. For his research he was elected a Fellow of the Royal Society in London.

Nearly a century would pass before Valvasor's example was emulated by another investigator. In 1747 court officials in Vienna ordered a 30-year-old mathematician named Joseph Nagel to explore and map the major caves of the Austro-Hungarian Empire. The reasons for the assignment remain obscure; conceivably Nagel suggested it himself. In the course of a two-year study, Nagel, accompanied by an Italian artist, mapped and sketched several Slovenian caves. His reports were unpublished, and reposed in obscurity in the royal archives. But while they did little to advance cave science, they apparently won the Emperor's favor, launching Nagel on a brilliant career. He soon was appointed court mathematician, then keeper of the royal scientific collections and head of physical science at the University of Vienna.

Both Valvasor and Nagel devoted much of their attention to the most famous Slovenian cavern, Adelsberg Cave (now called Postojna Jama by the Yugoslav government, which annexed the Slovenian Karst after World War II). Adelsberg's conspicuous entrance, a gaping tunnel with the sparkling Pivka River gushing from its mouth, attracted travelers as early as the 13th Century; one alleged visitor was the Italian poet Dante, who may have been seeking material for *The Inferno.* Vacationing nobles and landowners toured the cave occasionally, and their need for guides, candles and torches spawned a cottage industry for local peasants.

A nattily dressed Joseph Nagel points out a feature of a Moravian cave to two members of his expedition. Nagel's scrupulous documentation of his explorations of eastern European caves in the 1740s included drawings such as this one, made by an expedition artist, Carlo Beduzzi.

The cave's early fame was attributable more to its convenient location near the main thoroughfare between Vienna and the Adriatic Sea than to its speleological wonders. Its chamber contained a few soot-blackened stalagmites, and the passage seemed to dead-end after some 340 yards, blocked by rock walls and the impassable river. But in April 1818, workmen who were preparing for a visit by the Habsburg Emperor Francis I of Austria discovered a larger, fantastically ornamented chamber with an entrance high on the original cave's wall *(pages 42-49).*

There is no evidence that the 50-year-old Emperor made the climb to view this new marvel, but a district official named Josip Jeršinovič soon rendered it accessible to all and sundry—for a price. Jeršinovič was among the first to realize that fashionable people would pay to tour a spectacular cave, despite their fears, provided that the path was smooth and well lighted. After installing himself as the first manager and treasurer of Adelsberg Cave, he hired laborers to bridge the river, construct a stairway to give easy access to the newly discovered continuation of the cave and level a path through its two and a half miles of passages. To protect Adelsberg from the depredations of souvenir hounds—and its promoters from loss of profits—he created a commission that regulated every aspect of the cave's operation. A locked gate was installed at the entrance and torches were banned in favor of clean-burning candles and oil lamps. Visitors had to sign a logbook, buy tickets to enter, and hire specially trained guides.

Amateur geologists occasionally visited the cave, but the first precise survey and map were not made until 1821. And the farther dimensions of the cave system remained unknown until Adolf Schmidl, a 48-year-old professor at Vienna's Imperial Academy of Science, began exploring the Slovenian caves in 1850. Schmidl was both a daring explorer and a formidable intellect, schooled in philosophy and law as well as geology. He ventured along underground rivers in a specially made wooden canoe, which could be dragged through shallow water or even dismantled and carried through dry passages.

On the afternoon of August 30, 1850, when the Pivka River was unusually low, Schmidl and his son launched their tiny craft and entered Adelsberg. For most of the night they paddled upstream, scraping through two low-ceilinged passages that ordinarily were impassable sumps. Meanwhile, an evening thunderstorm had drenched the surrounding area, and at about 1 a.m. the river, swollen by runoff, rapidly rose nine feet, sealing the sumps and stranding the two explorers. For several anxious hours the pair waited in the clammy darkness, 1,800 feet inside the cave, conserving their supply of candles and lamp oil; when the river fell, they quickly departed. Later, aided by a professional surveyor, Schmidl prepared a precise map of the newly discovered river passage and meticulously reported on the subterranean flow of air and water.

But the primary interest in Adelsberg Cave remained commercial. After a railway line was built from Trieste to Adelsberg in 1857, special excursion trains made regular trips to the cave. By 1872 the cave commission was installing its own narrow-gauge railway inside the cavern, although at first

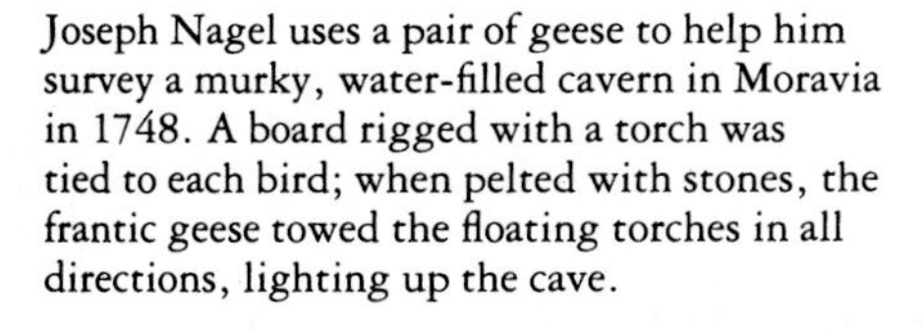

Joseph Nagel uses a pair of geese to help him survey a murky, water-filled cavern in Moravia in 1748. A board rigged with a torch was tied to each bird; when pelted with stones, the frantic geese towed the floating torches in all directions, lighting up the cave.

teams of guides had to push the little two-seat cars themselves, trundling the tourists through the cave like baggage in handcarts.

Some visitors thought the commercialization was overdone. "In the presence of this splendid freak of nature," wrote a *New York Times* correspondent in 1881, "it is somewhat disenchanting to find that the whole thing has been reduced to a system, and that you are confronted by a neat little wooden bureau, where a courteous old gentleman meets you with an inquiry as to 'which kind of illumination you wish to order.' Meanwhile your landlord, with a steadfast eye to business, essays to tempt you by sowing the tablecloth with photographs of the cave and its surroundings."

For most visitors, however, the cave's dazzling variety of stalagmites, stalactites and crystal formations overwhelmed such complaints. And even the *New York Times* correspondent was mesmerized by the glistening stone galleries—"carved battlements, wide-mouthed gargoyles, slender Moorish

In this 1689 engraving, the Austrian explorer Baron Johann Valvasor depicted the stalagmites and dripstone formations of a large chamber inside Adelsberg Cave as hideous gargoyles and fiendish monsters. Valvasor explored about 70 Slovenian caves, portraying each in the same fanciful terms.

columns, grim low-browed arches, fretted roofs, somber Gothic gateways, intertwined spirals, massive pillars festooned with cypress or palm leaves, tomb-like crypts, colossal chandeliers of cold white stone glittering diamond-like with countless drops of water." There is "something indescribably weird and unearthly," he wrote, in "the hollow roar of the Stygian stream below, the ghostly glimmer of its half-seen waters, the mighty void of the sunless cavern which looks all the vaster for those tiny specks of light which struggle in vain against the gloom of this shadow of death."

In America, attitudes toward caves were generally less romantic and more pragmatic. For thousands of years, Indians in the Kentucky karst belt, the most extensive cave region in America, had mined gypsum, a mineral found in cave formations, and used it to make ceremonial paints. Settlers on the rugged western frontier had an even more intense interest in caves: By the late 18th Century they had learned how to extract saltpeter, a vital ingredient of gunpowder, from nitrate-laden cave soil, which had been enriched by the guano from generations of bats.

According to legend, in the 1790s a hunter named John Houchin tracked a wounded bear through the wooded hills of central Kentucky to a large pit near the Green River. After dispatching the animal he entered its lair and found a huge chamber, its walls black with bats and its floor deep in nitrate-rich soil, recognizable by the ammonia stench of guano. A local speculator named Valentine Simons learned of the promising find and, for $80, bought a 200-acre parcel containing the cave entrance.

As war with England loomed in 1812, the threat of a blockade by the British Navy boosted demand for domestic saltpeter. The cave property changed hands several times, its value appreciating sixfold. One owner, boasting that the cave "could supply the whole globe with saltpeter," changed its name to Mammoth Cave. It would become as famous in America as Adelsberg was in Europe.

During the War of 1812 more than 70 slaves mined the cave's chemical wealth, shoveling soil onto oxcarts, hauling it to the cave mouth, dissolving the nitrate in wooden leaching vats and crystallizing it in boilers. The slaves and their owners had little time and less inclination to investigate beyond the nitrate deposits; these lay no more than 1,800 feet from the cave mouth—within easy flying distance of the outside world, where the bats made nocturnal forays for insects. By war's end, the miners had gouged huge holes in the cave deposits and had produced an estimated 400,000 pounds of saltpeter; but soon afterward gunpowder makers found cheaper saltpeter supplies overseas, and the mining operations tapered off.

The miners provided the cave with a curious new attraction, however. While working a nitrate cave about four miles from Mammoth, they had unearthed a mummy of an Indian woman, sitting upright in a stone coffin buried 10 feet below the cave floor amid an assortment of beads, feathers, snake rattles, a deer-foot talisman and utensils. The mummy was nearly six feet tall yet weighed only 14 pounds, apparently because of the desiccating effect of the guano. One of Mammoth's owners, an amateur historian named Charles Wilkins, learned of the discovery and arranged to take custody of the mummy—nicknamed Fawn Hoof for the talisman found in the crypt—and display it in the cave.

Fawn Hoof had been a sort of macabre guest in Mammoth for nearly a year when she was seen in 1815 by an Ohio physician named Nahum Ward, who immediately perceived exciting commercial possibilities. Ward arranged to exhibit the mummy in Lexington, Kentucky, and the shriveled apparition drew such crowds that he went on to travel the Eastern Seaboard, displaying the "Mammoth Cave Mummy" in Philadelphia, Boston and other cities. He also wrote a promotional article, embellished with a map of the cave and an engraving of the departed Indian woman, that was reprinted in dozens of magazines, broadsides and books in the United States and abroad. Before long, Ward's strenuous promotions had earned Mammoth Cave a place in an 1821 volume titled *The Hundred Wonders of the World.* Eventually the mummy was dispatched to the Smithsonian Institution.

Although the tour had made Dr. Ward wealthy, Mammoth still received a mere trickle of visitors. Its fortunes changed, however, after the cave and the land around it were sold in 1838 to Franklin Gorin, a lawyer from nearby Glasgow Junction. The turnabout was due less to Gorin's management than to the special gifts of Stephen Bishop, the 17-year-old slave he introduced as a tour guide.

Athletic, resourceful and fascinated by the countless surprises that the cave yielded, Bishop—always called Stephen—became one of America's most renowned underground explorers. Travel writer Bayard Taylor described him as "a slight, graceful, and very handsome mulatto with perfectly regular and clearly chiselled features, a keen, dark eye, and glossy hair and moustache. He is the model of a guide—quick, daring, enthusiastic, persevering, with a lively appreciation of the wonders he shows." Stephen, Taylor wrote, could discuss geography, history, literature and Greek mythology as well as the geology of his domain. "No one can travel under his guidance," Taylor continued, "without being interested in the man, and associating him in memory with the realm over which he is chief ruler." Such well-publicized encomiums soon swelled the tourist trade.

Dressed in his guide's outfit—a brown slouch hat, jacket and striped pants—Stephen led lamp-toting sightseers through a chamber called the Rotunda, a passage known as the Little Bat Avenue for its winter residents, an 80-foot-high side gallery called the Methodist Church and other fantasies in stone with similarly evocative names.

In his idle hours Stephen began to explore passages branching off from the main route. Penetrating a narrow aperture behind a large flat-topped rock called Giant's Coffin, he edged through a maze of virgin passages and found pieces of a cane torch left behind by Indian gypsum miners centuries before; nearby he found an immense, unfathomable pit. Stephen returned with owner Gorin and another man, who lowered him 90 feet to the bottom of what they christened Gorin's Dome. Such pits eventually led to dripping passages that contained larger, more colorful rock formations than the dry upper cave.

When an adventurous visitor from Kentucky named H. C. Stevenson asked to see an unexplored portion of the cave, Stephen led him through the jagged, twisting passages beneath Giant's Coffin, and on past Gorin's Dome to the Bottomless Pit, a six-foot-wide shaft whose opening entirely blocked the main corridor. Stevenson dropped a rock into the void and counted two and a half seconds before it hit. Then he and Stephen gingerly extended across the abyss a wooden ladder that Stephen had stowed nearby. Holding a flickering oil lamp in one hand, each man in turn straddled the rickety bridge and edged across the abyss. They jumped over a smaller hole, then duck walked down a low-ceilinged oval stoopway to a high, sand-floored passage. There, Stephen let his excited client become the first man to plant his footprints. A few more yards brought them to the brink of a canyon; below, they saw a sizable river—the first major watercourse discovered in Mammoth. Before turning back, they named it the River Styx.

In this way Stephen continually extended the cave frontier, squeezing through tortuous openings barely large enough for his lithe frame and emerging into chambers no man had ever seen. In his first year Stephen doubled the known length of the cave—and, equally important, generated a flurry of newspaper stories about the cave's receding frontier and its intrepid guide.

Unlike the notoriety of Fawn Hoof, the new publicity generated a long-awaited surge of tourism. Suddenly prosperous, Gorin bought two more slave guides for Stephen to train, built a cedar bridge over the Bottomless Pit and introduced a skiff that ferried sightseers along the Styx. Stephen himself became a major attraction, in part because a slave was not expected to be so intelligent and well-spoken. Tales of Stephen's strength and courage became part of the cave's growing lore. He carried one exhausted tourist for six miles and tracked down another who had strolled off alone, lost his lamp, knocked his head on the rocks and lain unconscious for 43 hours.

Within the year the flood of tourists overwhelmed the small log structure that Gorin used as an inn. Lacking the capital to expand, he sold the whole enterprise—cave, land, slaves and inn—to a wealthy Louisville doctor and gentleman-farmer named John Croghan in the fall of 1839.

Working for this new master, Stephen made yet another splendid discovery, seemingly on a whim, by climbing a slope above the river and following a long alley to a boulder-strewn choke. He and a visitor clawed away rocks with their hands, not knowing whether they would find a solid rock wall or another chamber. After hours of work they reached an opening and wriggled through, emerging on a ledge midway up the sloping wall of the deepest pit Stephen had yet seen. As they cautiously descended, their lamps illuminated rock ornaments rivaling those at Adelsberg. The ceiling of this great shaft, soon named Mammoth Dome, was invisible in the blackness 192 feet above the floor. In subsequent months Stephen explored a second stream, the Echo River, and opened up several more miles of broad boulevards, one of them leading to an enchanting chamber festooned with gypsum flowers and snowball-like growths on the walls—the Snowball Room.

To publicize Stephen's discoveries, which had long since outdated the 1835 map, Croghan resolved to print new maps and a lavish guidebook,

This engraving is one of the few pictures of the legendary slave guide of Mammoth Cave, Stephen Bishop. When Bishop died in 1857 at the age of 36, his talents as an explorer and self-taught geologist had already made him almost as famous as the cave itself.

A 19th Century engraving shows a party of tourists being paddled along Echo River, one of the largest underground streams in Mammoth Cave. While one stocking-capped guide regales an enthralled female passenger, another blows a horn to demonstrate the cave's wondrous echo effects.

"bound in neat half morocco with cloth sides, lettered and filletted on the back," according to the printer's contract. He set Stephen to work drawing a pencil map from memory during January 1842. Within a week the uneducated slave had produced an elaborate freehand sketch showing the cave's hundreds of sinuous passages, all drawn to scale—a document whose usefulness endured for 40 years.

Croghan enlarged the cave hotel from a two-room log cabin into a luxurious complex consisting of four main buildings and 16 cottages. He also persuaded the state legislature to help rebuild the wretched, nearly impassable roads between the cave and the main Louisville-Nashville stage route. Soon travelers were paying more for accommodations and meals than for the various cave tours, which ranged in length from two to nine miles. But Dr. Croghan had another, purely philanthropic project in mind. For years he had believed that the nitrates, pure air and constant temperature within caves might cure consumption, the 19th Century scourge that is now called tuberculosis. Several other physicians shared this conviction, founding their belief on reports that saltpeter miners had enjoyed unusually good health in the sunless atmosphere, and that a walk through such a cave, whatever the distance, caused surprisingly little fatigue.

Croghan decided to test the thesis. In 1842 he built a dozen cottages about a quarter mile inside the cave and accepted his first patient, a medical man who had diagnosed his own ailment as pulmonary consumption. After five weeks the physician departed, pronouncing himself "very much relieved." News of the apparent cure prompted several other doctors to refer patients to Croghan, but the initial enthusiasm quickly waned. Several patients left after short stays produced no improvement in their health, and two men died in the cave, withering like the shrubs the patients had planted in a pathetic attempt to embroider their bleak surroundings. Within a year the experiment was abandoned entirely. More than a decade later, Bayard Taylor shuddered at the sight of the would-be sanatorium: "The idea of a company of lank, cadaverous invalids wandering about in the awful gloom and silence, broken only by their hollow coughs—doubly hollow and sepulchral there—is terrible."

Mammoth Cave and its celebrated guide continued to thrive as a tourist attraction, however, its popularity eventually rivaling Niagara Falls. By

1850 the cave boasted 226 avenues, 47 domes, 23 pits and 8 waterfalls, most of them easily accessible from paths blasted into the rock by Croghan's workmen. The cave hotel became an exclusive summer resort for wealthy Louisville families and a standard stop for visitors to the area. The renowned Norwegian violinist Ole Bull gave a recital inside the cave, in a room known from then on as Ole Bull's Concert Hall; soprano Jenny Lind, the "Swedish Nightingale," visited a few years later.

In 1849 Croghan died—ironically, of consumption. His will left the cave to his family and stipulated that Stephen be freed seven years later. Once free, Stephen planned to buy the freedom of his wife and son and move to the recently established African republic of Liberia, but his dream was never realized: He died in 1857 at the age of 36, one year after becoming a free man. Years later Franklin Gorin, Stephen's former master, eulogized him aptly: "Stephen was a self-educated man. He had a fine genius, a great fund of wit and humor, some little knowledge of Latin and Greek and much knowledge of geology, but his great talent was a knowledge of man."

The Civil War put a stop to tourism, and business was slow to recover in the postwar period. Mammoth's owners worked hard—if not always ethically—to stimulate the trade. In 1875 another female mummy, discovered at nearby Salts Cave, was advertised as the original Fawn Hoof and exhibited in Mammoth. And Mammoth guides continued to arrange crowd-pleasing shows for visitors, good-naturedly tweaking the tourists' uneasiness. "A nervous person," a sightseer from New York wrote, "imagines great eyes staring at him from every corner, and cold hands, ready to grasp his shoulder, at every step."

Mammoth no longer had the field to itself. Rival showplaces like Wyandotte Cave in Indiana were also doing good business, in part because the nation's growing network of railroads made travel so much easier. In the eastern United States, the most resounding success was experienced—for a time—by three denizens of Virginia's Shenandoah Valley: Benton Stebbins, a struggling, 53-year-old photographer; Andrew Campbell, a 42-year-old tinsmith; and Billy Campbell, Andrew's 26-year-old nephew. Knowing that a railroad would be built across the valley, and that there were several small caves nearby, the opportunistic trio spent so much time

Tourists in Kentucky's Mammoth Cave share a novel underground picnic with their two black guides in 1892. By this time, the cave's facilities had deteriorated so much that tourism had fallen sharply from its pre-Civil War peak of 40,000 visitors a year.

A pile of flat stones emplaced by visitors from Kentucky reaches Mammoth Cave's 12-foot-high ceiling. Tourists were encouraged to leave such evidence of their visits until the National Park Service forbade the practice in 1941.

running down cave rumors and probing the valley's many shallow pits that they became known locally as "phantom chasers."

On the steamy morning of August 13, 1878, as they stopped to rest in a clump of brush on the side of sinkhole-pocked Cave Hill, an aptly named ridge about a mile from the town of Luray, they stumbled on what seemed to be a large rabbit hole at the bottom of a 10-foot-deep sink. As they slid into the sink they felt a cool breeze whispering up from the hole. After fetching tools and rigging a block and tackle to lift dirt and rocks out of the sink, the men labored with picks and shovels for days, enlarging the opening until Andrew Campbell, the smallest of the three, could slip through. Holding on to a rope, he slid down a muddy slope, lighted a candle and found himself in a narrow rift about 15 feet long and five feet wide, with a small hole at one end. He let go of the rope, squeezed through the hole, and stood rapt in a large, glittering, circular chamber about 10 feet across and perhaps 40 feet high. A slender white column in front of the hole reached from the floor to the ceiling, while, overhead, hundreds of sword-shaped stalactites reflected the light. A worried shout from his nephew Billy summoned him back to the surface.

Stebbins and the Campbells returned that night to explore further. The cave seemed to have an unlimited supply of wonders. One passage wound through massive stalagmites and delicate stalactites for several hundred yards; another led to a chamber roofed with soda-straw formations, which glistened bright red and white and tan in the candle glow. Andrew Campbell wandered along a broad, lofty boulevard, then stepped into a pool of water that filled the passage. He waded on in cold, waist-deep water, but began sinking in mud and slogged back out. His candle flame was still being blown back toward the cave's entrance, a sign of extensive passages ahead that could only be explored by boat.

Having found a splendid cave, the explorers now needed either to strike a bargain with the landowner who controlled the entrance or, better still, to buy the parcel themselves. Secrecy was essential. After emerging from the cave, they blocked its entrance with rocks, then changed their muddy clothes in a nearby field to hide the evidence of their nocturnal foray. The next day Billy Campbell quietly perused the courthouse records. The 28-acre parcel, once owned by a bankrupt merchant, had recently been sold at

Spectators throng Giant's Hall in Luray Caverns in Virginia during an 1878 "illumination"—a promotion that involved lighting the cave with hundreds of candles. For an additional fee, guests could dance on a specially laid wooden floor to the music of an orchestra.

A guide taps out a tune on 56 graduated columns of dripstone in Luray Caverns in Virginia. The formation, dubbed the Organ, was a popular stop on the early tours of the cave.

To get a better view of a curious formation called the Shipwreck, a visitor to Luray Caverns in the late 1800s holds up a tin reflector lantern that was provided by the cave's promoters.

auction for eight dollars an acre, but the buyer had so far failed to make the required $28 deposit. If he did not consummate the purchase, it would be auctioned off again at the beginning of September. This time it might go for as much as $20 an acre and require a $120 down payment—an enormous sum for three impecunious tradesmen.

During the next month the trio secretly raised $40 each. Stebbins obtained a loan by using as collateral his wife's organ, a prized possession that she had inherited from her parents; Andrew Campbell swore a Masonic brother to secrecy and asked him for a loan; Billy Campbell borrowed from his father, the Page County Sheriff. Then they waited to see whether the previous buyer would put down the deposit, all the while pretending to search for caves and never visiting their discovery.

The property went on the block on September 10. To conceal their interest, the explorers had Billy's father, a courthouse regular, represent them. Bidding was surprisingly hot. The bankrupt's family were buying up his estate, and their surrogate drove the price of this rocky, seemingly worthless tract to $17 an acre before yielding. The new owners immediately paid the $96.90 deposit, recorded the sale and began to celebrate. They were in business—or so they thought.

As they began to plan their show cave, the new owners worried that its chambers, while grand, might be too small to attract tourists. In mid-September they returned with a dinghy, enlarged the entrance to accommodate it and launched their little craft on the lake. They drifted soundlessly down the broad, watery avenue, passing under a sculpted rock bridge and nosing into shore at the base of a muddy slope. At its top, a long, level corridor led to a vast chamber walled with parti-colored flowstone draperies. Enormous columns and stalagmites met them at every turn; folds of orange and red curtained the walls. They clambered about for two hours in overjoyed awe before returning to the boat.

Stebbins and the Campbells fell to work readying the cavern for visitors—building stairways, bridges and handrails—and tourists began arriving almost immediately. An enthusiastic *New York Herald* account of the cave's discovery attracted visitors from the entire Eastern Seaboard and furnished a simple, catchy name—Luray Caverns. In November, Stebbins, a natural showman, placed thousands of candles throughout the cave and invited the public to a "grand illumination" at 50 cents per adult and 25 cents for children, netting $91 profit in one day. The overnight success of the phantom chasers prompted hordes of men to scour the countryside in search of comparable bonanzas. "There is not a promising rat hole in Page County," the *New York Herald's* man observed, "that has not been 'opened up' in the hope that it might lead to great cavernous spaces below."

But the good fortune of Stebbins and the two Campbells was short-lived. Early in 1879, relatives of the former owner of the cave property, seeking to save the rest of his far-flung estate from the auction block, filed a lawsuit charging that the group had defrauded them by concealing knowledge of the cave at the original sale. Until the suit was settled, no bank would loan money for desperately needed improvements—boats, bridges, footpaths and the like. As the case dragged through the courts, Stebbins arranged to sell the property to the Shenandoah Valley Railroad Company for $40,000, enough to make all three partners rich for life—but they could not consummate the sale until they won the lawsuit.

The circuit court found in favor of the three explorers, but on April 21, 1881, the Virginia Supreme Court unanimously ruled against them. The original owner's family reclaimed the cave, then turned around and sold it to the railroad for $39,400. Under its new operators, Luray Caverns would become one of the best-known and most profitable caves in America. Stebbins and the Campbells got nothing, although Billy Campbell was subsequently employed as a cave guide.

Just before the hapless threesome lost the cave, they welcomed a scholarly delegation from the Smithsonian Institution to Luray. The report of this august panel was a marvel of fevered imagination that neatly illustrated the popular attitude toward caves in 1880: "Here in this dark studio of nature are reproductions of all those objects which fill the mind with pleasure, wonder or alarm," among them "spectral beings—terrestrial, celestial and infernal." The prose was strikingly similar to the contemporary *New York Times* article about Adelsberg Cave, which dwelled on "the hollow roar of the Stygian stream below, the ghostly glimmer of its half-seen waters," and "the gloom of this shadow of death." Such romantic accounts did not consider how these caves had come into being, or what processes accounted for their spectacular formations and exotic inhabitants. Such questions—and their answers—awaited the birth of a new kind of science in the next century. Ω

The first cavern to achieve worldwide renown as a showplace was Adelsber Cave, located near a village of the sam name in Austria. Explorers began prob ing it as early as the 13th Century, bu the cave received only modest attentio until the Holy Roman Emperor Francis decided in 1818 to make a firsthand in spection of its reputed wonders.

The royal visit was preceded by a flur ry of activity in the cave as worker cleared away rubble and climbed ricket ladders to install torches. In the process one laborer happened upon a gap in th cave's far wall, some 90 feet above th floor. Crawling through the opening he found a new section far larger tha the first, bedecked with fantastic drip stone formations.

The Emperor did not venture int this portion of Adelsberg, but local offi cials reasoned that, if problems of ac cess could be conquered, the publi would find it irresistible. At their be hest, workers leveled paths through th chambers, built a wooden bridge acros the cave's underground river and chis eled stone stairways into the walls *(over leaf)*. The crude torches were replace with chandeliers and oil lamps, an guides were trained. To illustrate a de luxe guidebook, an artist named Aloi Schaffenrath executed 19 watercolors c the cave; a sampling appears here and o the following pages.

Adelsberg was soon drawing 1,00 visitors a year—and that figure wa quickly surpassed after a railway lin was opened to Adelsberg in 1857. B the 1870s, the annual number of tour ists had soared to 8,000 and a fashion able hotel had been built nearby. Eve Americans, amply endowed with cave of their own, were forced to acknow edge the appeal of this European gem.

After visiting Adelsberg in 1881, reporter for *The New York Times* wrote i awe of the "stalactites of unequale splendor" and the "fantastic architec ture" in the cave. Indeed, concluded th *Times* man, although Adelsberg lack "the mighty expanse of the Mammot Cave of Kentucky, every part of it filled with a stern and gloomy grandeu which is indescribably impressive."

In a watercolor of Adelsberg's exterior by Alois Schaffenrath, tourists line up to enter the cave by way of a hillside portal. Below, the Pivka River emerges from another opening.

Inside Adelsberg's Great Hall, visitors pass between two upper galleries by descending a staircase, crossing a bridge over the Pivka River and climbing a second staircase on the opposite side. A chandelier, oil lamps and the guides' torches light up the dripstone formations of the 115-foot-high vault.

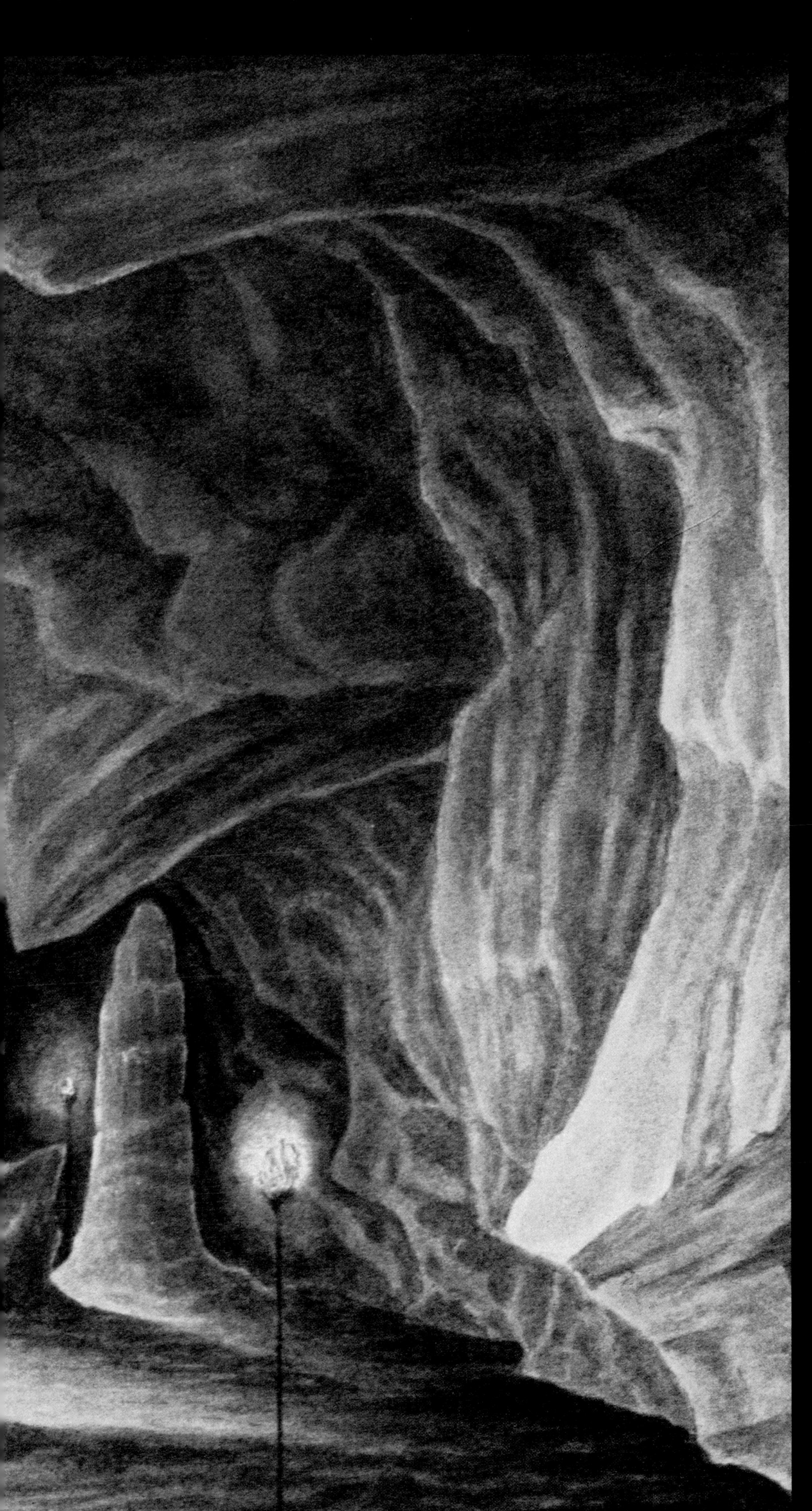

Three tourists admire a nearly transparent stone drapery known as the Curtain. As in many of his cave views, the artist whimsically depicted himself sketching the scene.

Their way barred by fallen columns, tourists prepare to leave Adelsberg Cave on a well-constructed path. In this painting, Schaffenrath included not only himself *(lower left)* but three assistants who are illuminating the far reaches of the chamber and his sketch pad.

Most of the world's caves were formed over hundreds of thousands of years by the slow dissolution of limestone, but some spectacular caverns are the result of processes that can be completed in a matter of days or weeks.

The lava that spews from a volcano sometimes cools and solidifies rapidly on the surface while remaining fluid underneath. Subsurface streams of molten rock may then course through slowly hardening channels in the lava flow and drain away, leaving a network of tunnels behind. The largest known lava cave, Kazimura Cave in Hawaii, extends for more than seven miles.

Glacial ice is another medium for cave making: Hollowed out by meltwater and warm winds, glacier caves are perhaps the most beautiful of all. The dense ice filters the light into brilliant shades of blue, and the wildly sculpted walls reflect color like the facets of a gemstone. But these caves are also the most dangerous: Melting ice is inherently unstable; roof collapses are frequent; sudden floods of icy meltwater are common.

In arid climates, dust particles driven by powerful winds scour vulnerable strata in cliff walls and create large, shallow caves. Often perched high above the surrounding terrain, wind-carved caves provided safe and enduring shelter for ancient communities in the Upper Nile Valley and in the American Southwest.

Less suitable for permanent human habitation are sea caves, coastal caverns excavated by ceaseless wave action. Over the millennia, waves blasting into weaknesses in rock formations have gouged out thousands of spacious caverns along the seashores of the world.

The blackened, rugged walls of two cave passages in the Lava Bed National Monument in California commemorate an ancient stream of molten rock that diverged into two channels. Located near a shield volcano called Medicine Lake Highlands, these tunnels are part of a lava cave network that underlies 72 square miles.

Pounded by the waters of the southern Indian Ocean, sea caves perforate the shoreline at Port Davey, Tasmania. Some of these caves extend as much as 200 feet inland.

Dressed for a cold, wet tour, a visitor inspects a subglacial waterfall inside a deteriorating remnant of Muir Glacier in Alaska. Because of rapid melting, glacier caves frequently change their shape from week to week.

Safely ensconced in the side of a sandstone cliff, the ruins of an Indian village—now called Cliff Palace—fill a cave carved by water and wind at Mesa Verde in Colorado. This village was built in the 13th Century.

local caves, villagers directed him to an open field that at first glance seemed unremarkable. But at its center was a gaping round pothole 115 feet in diameter—the unexplored pit of Padirac, which had figured in local folklore for centuries. When Joan of Arc drove the English from France, according to legend, several British soldiers took refuge in the cave and hid an enormous sum of gold there—a hoard that never had been recovered.

By dropping his sounding line all around the lip, Martel learned that the pit was 247 feet deep at one side but only 182 feet deep at the opposite edge—presumably because of a steep mound of fallen rock. At about noon, after his crew had unloaded the horse-drawn wagon and had arranged the rigging, Martel descended by the shortest route, climbing down a 180-rung rope ladder to the rock pile. He was followed by his companions, Gaupillat, Foulquier and Armand. As he looked up at the entrance, he observed that he seemed to be "at the lower end of a telescope, with a circular bit of blue sky for object-glass. Around the margin of this enormous cylinder hang long, graceful streamers of plants that love shade and dampness. Our slender telephone-wire looks like a dark spider's thread across the chasm."

As they scrambled down the jagged rock pile, Martel and his companions discovered that the north and south sides of the sloping mound extended beneath the overhanging walls of the bell-shaped pit, leading to two passages: a gently sloping gallery with a 65-foot-high ceiling on the south, and a steep, narrow crevice on the north. Leaving the twilight of the pit, they walked downhill by candlelight into the lofty south gallery, following the musical rippling of a hidden rivulet. After going 300 feet, they found a sink where the stream disappeared beneath the rock mound. This rivulet, Martel noted as he traced it upstream, was flowing along a bed of impermeable shale that trapped rain water seeping down through the limestone. Although such shale formations sometimes presage magnificent caverns, the course of this little stream soon ended at a rocky choke point.

Returning to the pit, the explorers cautiously clambered down the steep, unpropitious north crevice. After about 300 feet this passage suddenly opened on a stately gallery 30 feet wide and as much as 130 feet high, where they found the downstream resurgence of the original stream, meandering among the stalagmites. As legions of bats fluttered about the roof, the explorers illuminated the chamber by burning magnesium wire but still could not see the end. They followed the widening stream for nearly a fifth of a mile, wading through several frigid pools, until their way was blocked by water several yards deep. Rather than launch his boat that evening, Martel prudently ordered a retreat at about 7 p.m. After a strenuous two-hour climb out of the pit, he and his crew bedded down for the night.

At six the next morning, July 10, the final day of Martel's summer vacation, the four explorers laboriously descended one by one, received the bags containing the collapsible boat and hauled them through the crevice to the water's edge. After assembling the fragile craft and rigging its lights—four candles on the gunwales and a magnesium reflector at the bow—Martel and Gaupillat embarked at 10 o'clock, leaving the other two to wait on shore. While one of the explorers slowly paddled downstream through the still water, the other kept watch in the bow, took notes and occasionally lowered a weighted line to measure the water depth—an extraordinary 16 to 26 feet. The only sound during this eerie voyage was the patter of water droplets falling from the gallery's invisible roof—"a melodious song, sweeter and more harmonious than the dulcet tones of the upper world," rhapsodized Martel.

After 1,100 feet the silent river gently rippled through a little rapid.

Martel and Gaupillat waded ashore, carried the dripping boat over the rocks, and halted in amazement. Below the rapid, the narrow passage opened into a chamber perhaps 200 feet long and 50 feet wide, containing four pristine lakelets and a profusion of intricate dripstones, glittering in the brilliant white light of the burning magnesium. "Detached columns, pendants, girandoles, more than sixty feet long, hang from the ceiling to the surface of the water," Martel later wrote. "Along the sides rise tiers of crystal bouquets, fonts, statuettes and spires."

The explorers cruised easily through the lakelets, but beyond them lay a formidable series of obstacles that demanded stamina and resourcefulness. Below the lakelets they encountered the first of a series of semicircular rimstone dams, each impounding a pool as much as 26 feet in diameter and 13 feet deep. At each one they stepped onto the delicate white crystals of the dam's inch-wide lip (a destructive practice modern cavers would eschew), held their candles in their teeth and lifted the canoe into the next pool—a dangerous maneuver during which they occasionally slipped and fell bodily into deep water, only to remount the dam and try again.

Beyond the second dam the cave roof abruptly lowered until it was barely a foot above the water. The explorers shipped their paddles, crouched below the gunwales and pushed through the 18-foot-long passage to the third pool, snapping off stalactites that barred their way. Next was another canyon, this one with a stream bed too narrow for the canoe. In waist-deep water the two had to lift the dripping boat overhead where the walls widened and grope forward until they reached a circular lake 150 feet wide, dotted with stalagmite islands. While Gaupillat illuminated the lake with the magnesium reflector, Martel edged carefully along the narrow ridges until he found an outlet—only to fall into the 57° F. water during his return. "The cold bath was not at all disagreeable," he insisted.

For hours the explorers voyaged steadily on into the fantastic cavern, past a series of 12 more rimstone dams and a 114-foot-wide lake encircled by sheer walls, until two thirds of a mile from the entrance pit they reached a canal only 28 inches wide—again too narrow for the canoe. After some deliberation they cautiously climbed the slippery flowstone walls and straddled the canal about five feet above the water, then raised the canoe with boat hooks and slid it along the rocks; its ribs creaked in protest, but the canvas held. Beyond the canal they found a lake and a hemispherical vault 65 feet high, devoid of stalactite ornaments. Seeing no exit, Martel at first assumed that the river finally disappeared in this stark chamber, but a careful circuit by boat revealed a wide tunnel barely 20 inches high at the far side. Crouching low, the two pushed through the tunnel for 32 feet to a soaring gallery containing the largest lake yet, a tranquil 250-foot-long pool that they named Lac de la Chapelle.

The explorers continued to press forward, hoping to reach the cavern's terminus; by 2 p.m. they were a half mile from the landbound foremen and a mile and a quarter from the pit. There, facing the 33rd rimstone dam of the day and the prospect of untold obstacles ahead, a weary Martel reluctantly decided to retreat, leaving further exploration for the following year. As he later recalled, "We are dripping wet; the chilly water is beginning to benumb our limbs. Our stock of candles is nearly exhausted." The return voyage proceeded uneventfully until the weary explorers tried to drag the canoe over the rough edge of a rimstone dam, tearing the canvas so that constant bailing was required. At 4:30 p.m. they finally reached Armand and Foulquier, who had anxiously begun to plan a rescue party, and by 7 o'clock all four men were sitting down to dinner on the surface, basking in a magnificent summer sunset.

From this day forward Martel was infatuated with Padirac, the most spectacular cave heretofore discovered in France—"a perfect fairyland for wonder and beauty," he called it. During his next campaign he explored the cave for 23 continuous hours, tracing its stream an additional 600 yards and discovering a spectacularly ornamented side grotto 260 feet high. In ensuing years Martel conducted a long struggle to open Padirac to the public, doggedly battling local landowners. Thanks to his efforts a philanthropic society was formed that bought the cave, built an enormous iron stairway into the pit, installed electric lights underground and, in 1899, opened Padirac to boatloads of tourists. Before selling, the former owner stipulated that he receive the legendary English gold if it was found—but nary a farthing ever turned up.

In subsequent campaigns, Martel explored hundreds of European caverns, including Adelsberg Cave, where he discovered more than a mile of virgin passage. The diminutive lawyer's international renown soon attracted notable patrons: Austria's Archduke Salvator asked Martel to investigate caverns on the Balearic Islands, Russia's Emperor Nicholas II asked him to explore the northwest coast of the Black Sea, and the French ministers of education and hygiene financed several expeditions to Austria and Belgium. But the crowning triumph of Martel's career occurred not on the Continent but in England.

In 1895 Martel decided to conduct his annual summer campaign in the British Isles, timing the expedition to coincide with the annual International Geographical Congress in London, where he was to lecture about cave-hunting methods. After exploring several caverns in Ireland and in the Peak District of Derbyshire, he journeyed to cave-pocked Yorkshire and Gaping Gill, a famous pothole that had frustrated British explorers for a half century. Plumb lines revealed the pit to be about 330 feet deep, shallower than

Paying out a telephone line for communication with observers on the surface, Édouard-Alfred Martel descends alone in 1895 into Gaping Gill, a 330-foot-deep pothole cave in Yorkshire, England. The descent took 23 minutes and left Martel soaking wet.

The river known as Fell Beck cascades 90 feet to the floor of a large chamber at the bottom of Gaping Gill. This engraving was based on a sketch made by Martel.

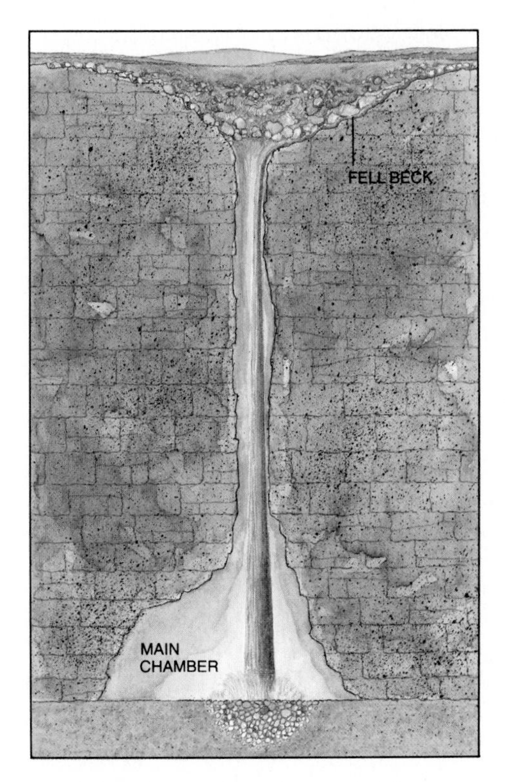

The Fell Beck drops through some 240 feet before it tumbles into the gallery at the base of Gaping Gill, as shown in this cross section. The great chamber was enlarged by the collapse of part of its ceiling and the subsequent dissolution of the rubble by the powerful waterfall.

some Martel already had conquered. The main obstacle, however, was Fell Beck, an icy stream that cascaded down the pit.

At Martel's request, the pit's generous owner, the lord of Ingleborough Manor, ordered workmen to reexcavate a 1,000-yard trench that had been used to divert Fell Beck from the pothole for previous exploring attempts. This three-day project was only partly successful, because heavy July rains had swelled the river, but Martel decided to proceed nonetheless. On the morning of August 1, he and his wife arrived at the pit, accompanied by a half-dozen workmen and about 80 spectators. A stout oak post was driven into the ground at the pit's edge while Martel, working without his vaunted French assistants, assembled his tackle and personally tested every knot. His three rope ladders totaled only 270 feet in length, so he tied them together and fastened them to the post with 60 feet of doubled hemp rope; the rope snaked down the pit's steep, funnel-shaped mouth while the ladder hung in the vertical shaft.

After three hours of meticulous preparation, Martel donned blue linen coveralls, filled his pockets with candles and magnesium wire, and strapped on his telephone while his wife (who never ventured underground) tested the surface telephone and instructed the men holding the safety line. At 1:22 p.m., a time he recorded with characteristic precision, he clambered down the doubled rope and began the descent, throwing loose stones down before him. As he climbed down the 13-foot-wide shaft, he rapidly was enveloped by the undiverted remnants of Fell Beck, "half-suffocating whirls of air and water" that doused him and filled his telephone. The cascade redoubled 130 feet down, where he had to descend through a frigid torrent gushing from a large fissure.

Half drowned by the waterfall, Martel gratefully paused on a six-foot-wide ledge 190 feet down, which had snagged the rope ladder. He untangled the huge heap of rope and dropped it into the narrowing pit, then threw himself back from the shower of rocks dislodged by the swaying ladder before continuing. Fifty feet below, the walls of the shaft suddenly receded and Martel found himself swinging like a pendulum near the roof of one of the largest caves yet discovered—"an immense Roman nave nearly 500 feet long, 80 feet wide and 90 feet high." Thirty feet above the floor he was forced to stop again, dangling tensely while another length of rope was added to his safety line. He alighted on the cave floor only 23 minutes after he began the descent, stepping six inches from the last rung of the ladder to the cave floor.

For an hour and a quarter Martel reveled in the spectacle inside the cave—and his triumph over France's traditional competitors, the English. "From the roof of this colossal cavern fell the cascade in a great nimbus of vapor and light," he wrote. "The most pleasant feature, however, was the thought that I had succeeded where the English had failed, and on their own ground." In the dim daylight that filtered down from overhead he paced off distances on the sandy, level floor, sketched maps and measured temperatures. At the south end he climbed a 60-foot pile of limestone boulders and at its summit heard the roar of an underground stream. "By clearing out these boulders it might be possible to discover other large caves," he conjectured, a prediction that eventually proved true.

At 3:30 p.m. Martel, soaked with water and chilled to the bone, prepared to ascend but found his telephone disabled by water; he could not hear his wife, although unbeknownst to him she could faintly hear his orders. After much shouting, he felt the safety line tighten and he laboriously began to climb the ladder, only to be halted 30 feet up when the safety line jammed. As he later recounted the predicament: "The waterfall

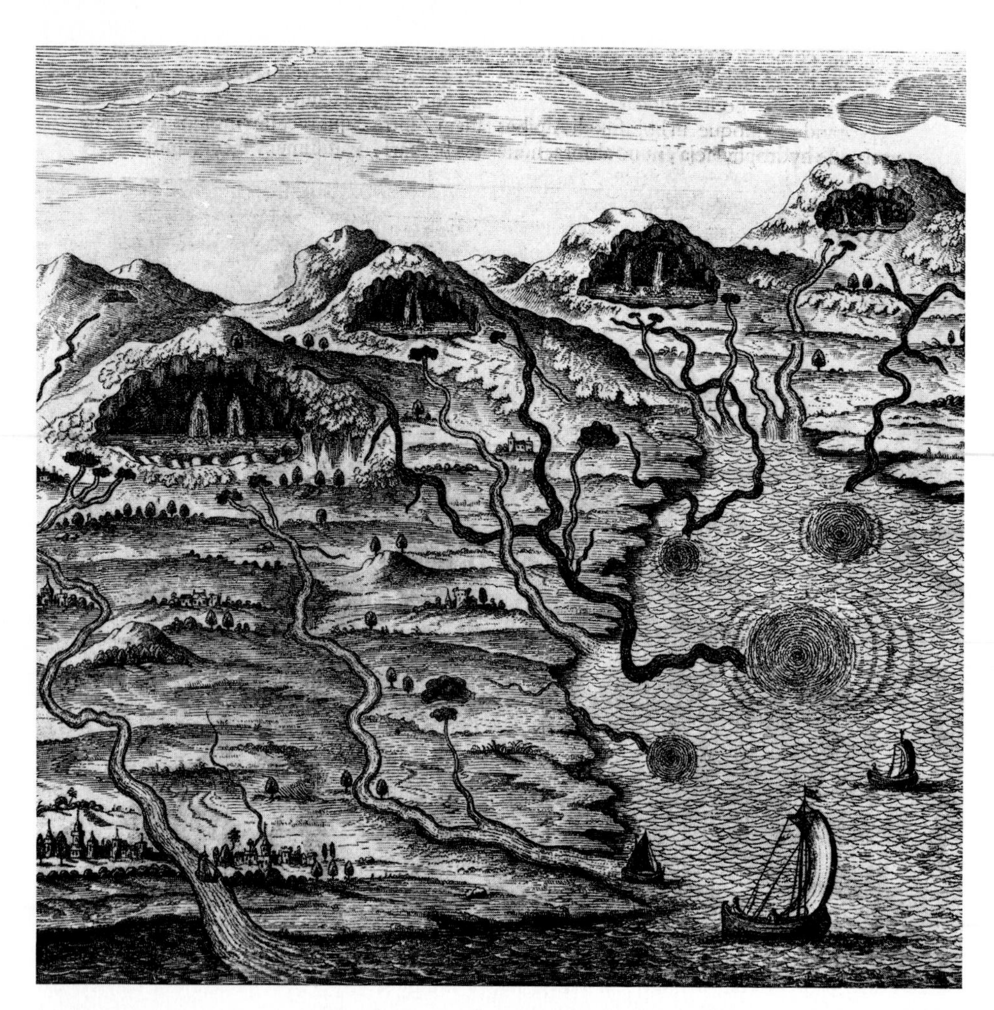

A 1665 illustration by a German Jesuit named Athanasius Kircher expounds an early—and erroneous—explanation for the presence of groundwater in caves. Kircher guessed that sea water was sucked down by whirlpools, then pushed by wind and tides through caves to vast subterranean reservoirs.

is freezing me. I try to climb the ladder without the life line, but my limbs will not carry me more than five rungs"—just enough to relieve the strain and free the line. Halfway up, his telephone wire was severed by a sharp rock, and his anxious wife could no longer make out any of his shouted instructions. Finally, after a strenuous 28-minute climb, he emerged from the pit blue and breathless from cold and exertion.

Martel continued to explore caves until 1914, mounting dozens of expeditions in France and abroad, but he increasingly concentrated on organizing and codifying the infant science of speleology. In 1895 he founded the French Société de Spéléologie, a pioneering group devoted not merely to exploration but to the scientific study of caves that soon spawned imitators elsewhere. A prolific author, he edited *Spelunca,* the society's bulletin, and wrote books about his own annual campaigns. Inevitably, Martel's cave-hunting avocation quickly overwhelmed his halfhearted law practice. In 1899 he abandoned law altogether and began to teach speleology at the Sorbonne. In his middle years, he became a professional speleologist, making a living from teaching, writing and consulting.

From his lectern Martel forcefully argued that speleology, which previously had been considered a sport or a singular eccentricity, should be recognized as a full-fledged science—"a subdivision of physical geography, like limnology for lakes and oceanography for seas." This viewpoint soon won reputable converts, as Martel's own career demonstrated. In 1904 he and his brother-in-law, Louis de Launay, a caving comrade, were appointed codirectors of *La Nature,* a French scientific journal. In 1907 Martel won the grand prize for physical sciences from the French Academy of Sciences, and in 1928 he was elected president of the Geographical Society of Paris.

As a scientist Martel was not particularly interested in abstruse theories about the chemical interaction of water and rock. His main venture into such abstractions was a rather eccentric warning that the earth was rapidly drying up. "The deepening and widening of rock crevices by water erosion and corrosion," he maintained, "causes water to descend lower and deeper into the bowels of the earth. Many centuries will not elapse before men will die of thirst and the earth itself perish of dryness." To avert this peril he recommended damming rivers, both on the surface and underground, and planting trees to diminish the infiltration of water into the earth.

But Martel was unsurpassed as an observer of the physical world—"the master class of lessons about *things,*" he called it. He amassed whole volumes of data about cave geology, hydrology, meteorology, flora and fauna, and he was a shrewd student of the practical problems of caving. Martel's greatest contribution was his research on how subterranean water circulates—a study prompted by his own bout with ptomaine poisoning, which he contracted from spring water in 1891. After recovering from the illness, he determined the spring's source by dumping fluorescein dye into a nearby stream sink, a hole where a surface stream plunged underground; shortly afterward the spring turned green. He then descended the pit and found the putrefying carcass of a dead calf that had contaminated the spring with what Martel wryly called "veal bouillon."

After further studies of several public springs, he proposed in 1892 a distinction between the water from "true springs," which was filtered by passage through fine sand, and that from "false springs," which merely flowed through fissured limestone. (The sand filters of true springs actually are not proof against bacteria, but true-spring water trickles through deep layers that rarely are exposed to contamination.) It would be impractical to make dye tests of every spring, so Martel proposed a temperature test based on his studies of cave meteorology. Since the air temperature in caves varies slightly but measurably according to the season, he reasoned, the temperature of water exposed to that air also should vary; by contrast, the temperature of true springs should remain as constant as the unchanging temperature of the surrounding earth. He recommended measuring water temperature four times a year and closing springs that varied by 1.8° F.

For several years Martel's repeated warnings about false springs fell on

After examining the spectacular stalactites and stalagmites in a cave on the Aegean island of Antiparos—shown here in an 18th Century illustration—French botanist Joseph Tournefort declared it "impossible this should be done by the Droppings of Water." He insisted that the rock formations were living vegetation and that they "grow from seed."

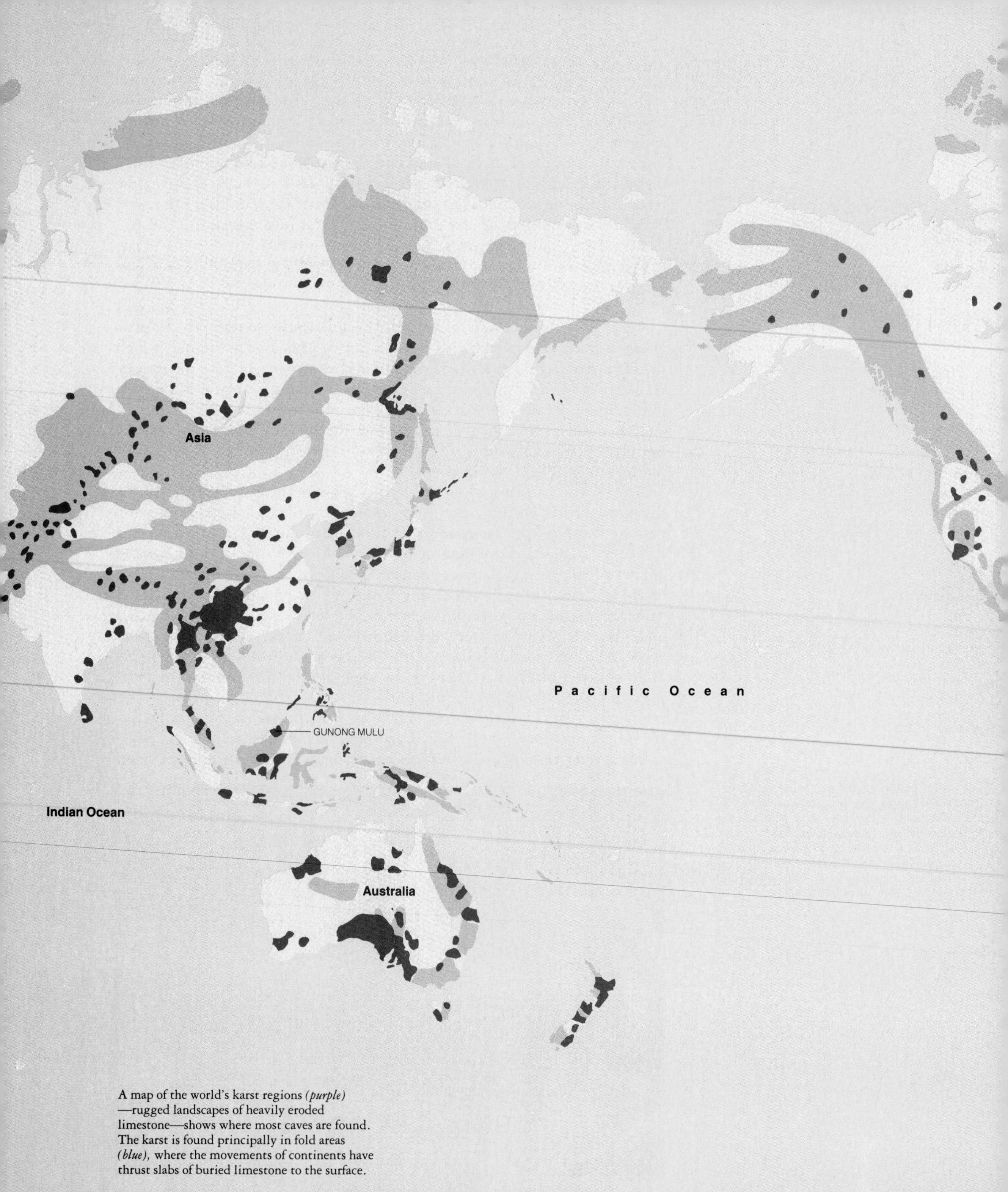

A map of the world's karst regions *(purple)* —rugged landscapes of heavily eroded limestone—shows where most caves are found. The karst is found principally in fold areas *(blue)*, where the movements of continents have thrust slabs of buried limestone to the surface.

The cave draperies of Carlsbad Cavern in New Mexico, seen here from underneath, were formed by water trickling down an inclined ceiling. The deposits began as winding trails of calcite, which were later extended downward in graceful curves by successive deposits.

Some of Carlsbad's banded draperies, stained with stripes of orange and yellow, resemble strips of bacon. The white portions are made up of relatively pure calcite; the rust-colored layers contain iron and other impurities.

Flat deposits called shelfstone build up along the edges of cave pools and around submerged columns at the water line. In the New Mexico cave seen here, the deposits mark the former level of a pool that has partially drained away.

Cave pearls glisten in a shallow pool in a cavern in New Mexico. As calcite built up around tiny particles in the pool, the constant agitation of dripping water rounded the pearls and kept them from sticking to one another or to the bottom of the pool.

In Angelica Cave in central Brazil, the repeated overflowing of a pool deposited calcite along the high-water marks, forming rimstone dams. Later, when the pool grew still, calcite precipitated on its surface as rounded, floating cave rafts. Each raft will eventually sink, either from its increased weight or because of a disturbance of the surface.

A gypsum flower in Kentucky's Mammoth Cave *(left)* is composed of crystals of calcium sulfate that precipitated on porous rock. Each "petal" was forced out into the cave as new crystals formed behind it.

Crystals of aragonite deposited by slowly seeping water in a cave in France create quill-like clusters known as frostwork *(right)*. Though chemically identical to crystals of calcite, aragonite crystals differ in shape and align themselves to form more irregular structures.

Extremely rare giant quartz crystals, as much as three inches in length, abound in this Arizona cave. Some geologists believe that such crystals form underwater in flooded chambers.

its startling resemblance to fried eggs to a principle of optics. The large crystals of the inner circle reflect yellow light; because the smaller crystals surrounding them have more facets per square inch, they reflect a broader range of colors, and appear white.

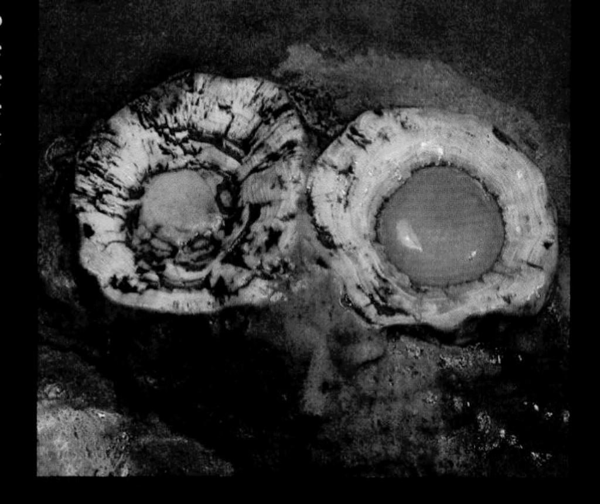

A structure that suggests the form of a butterfly, found in Sonora Caves in Texas, is the result of two kinds of water action. First, spiky crystals precipitated from seeping water; then dripping water added smooth layers of calcite over the angular framework.

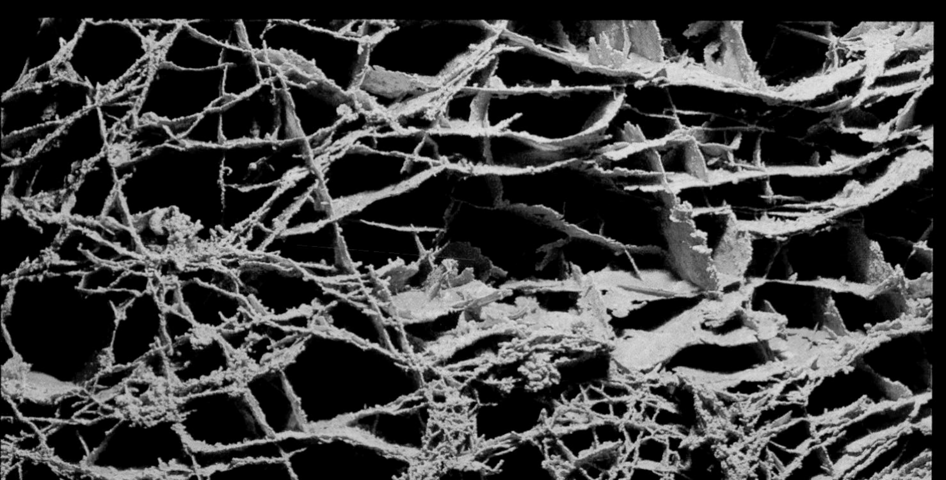

Deposited in cracks in the limestone of South Dakota's Wind Cave, hard crystals of calcite remain as a weblike formation called boxwork after the more soluble limestone eroded.

Tightly coiled helictites in a cave in New Mexico seem to defy gravity as they grow upward. Water rises slowly through their minute central canals by capillary action, then deposits crystals

CHEMINS DE FER P.O.. MIDI..P.L.M.
L'AVEN ARMAND
LA MERVEILLE SOUTERRAINE
Circuits d'Auto-Cars P.O.. Midi..P.L.M.
aux Gorges du Tarn et de la Jonte

hydrogen-sulfide brine rose through the earth from nearby oil and gas fields. The tremendous scale of Carlsbad's chambers, with their high, vaulted ceilings, is apparently the result of the acid eating its way upward over a period of thousands of centuries. About a million years ago the first stalagmites and stalactites began to form, a drop at a time in the dry air, and they are still growing.

As for Jim White, he served the National Park Service as its chief Carlsbad ranger until 1929, when he resigned from that post in the hope of being named "chief explorer," a role he deemed more fitting to his tastes and abilities. He was turned down, ostensibly because the Park Service director felt there was no need for such a position, since "already more rooms have been discovered and explored in this cavern than can now be taken care of or shown to the visiting public."

Shunted away from the cavern he had discovered and to which he had

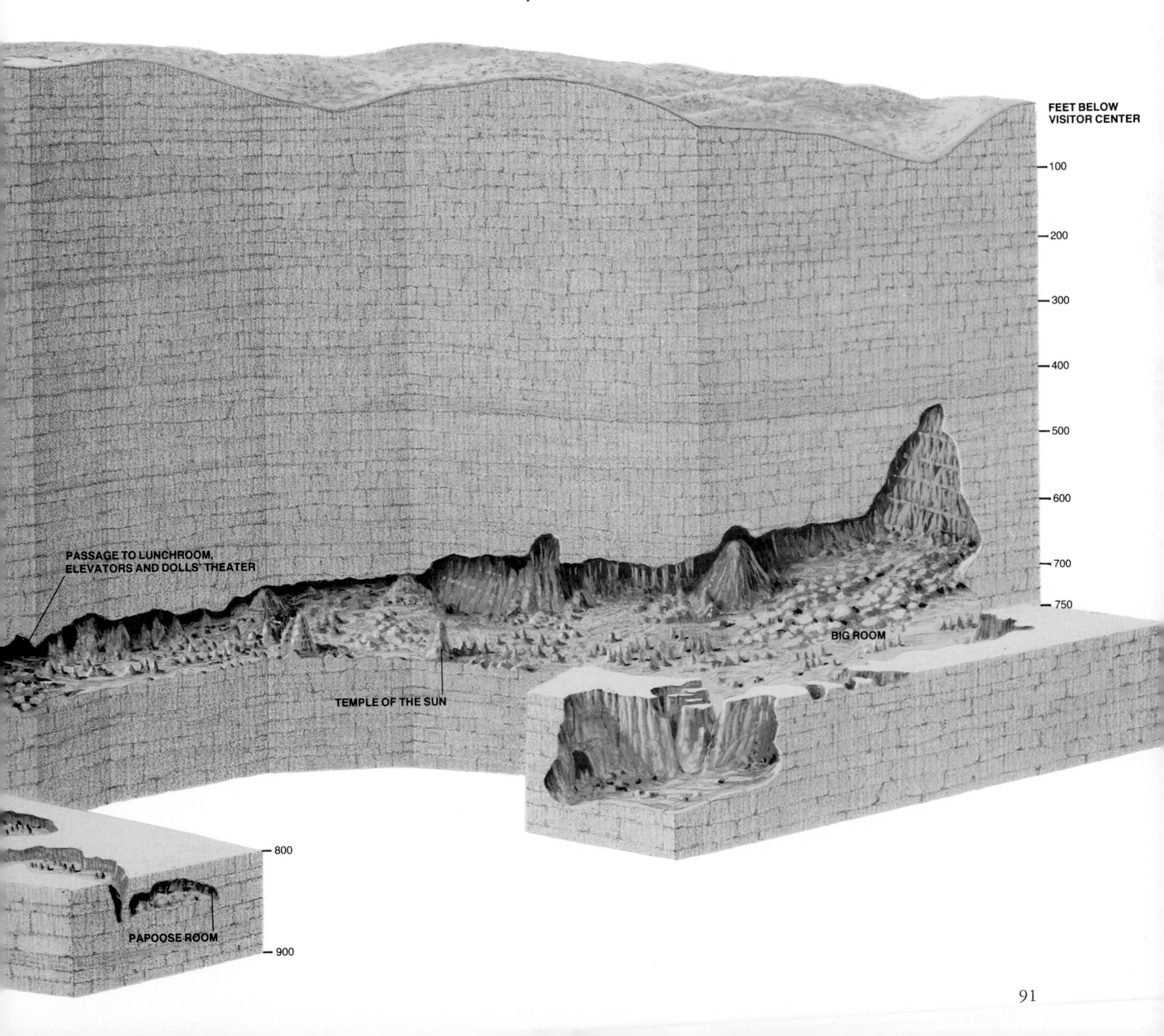

devoted so many years of his life, White died in 1946, an embittered man. Yet thanks largely to government sponsorship, his beloved cave at least was spared the indignities visited upon many other American caverns during the first several decades of the 20th Century.

During that turbulent time, the ageless caverns became the objects of greatly increased interest, not only in the United States but in Europe. The fascination took strikingly different forms.

In the United States, caves were generally exploited, either for their natural resources (as in the case of the guano mining at Carlsbad) or as commercial tourist attractions. Competition among rival cave entrepreneurs was intense and sometimes even violent. Disputes over ownership reached the highest courts in the land and, if not settled by legal decision, often resulted in bankruptcy or federal proprietorship.

In Europe, the tradition of speleology as the sporting science pioneered by Édouard-Alfred Martel was passed along to new generations—with the sporting aspects taking distinct precedence over the scientific. To the Europeans, the underground was a wonderworld of pristine beauty, of romance and of adventure, where death was to be dared if not defied. Speleological success was calculated not in terms of showplace profits but by the number of record-setting superlatives that could be applied to the discovery and exploration of caves.

Two men sum up the contrasts of the era. In Europe, the model of speleological endeavor was a scholarly Frenchman who had been attracted to caves by his boyhood reading of Jules Verne's fantastical *Journey to the Center of the Earth.* In America, the symbol of speleology was an almost illiterate Kentucky farm boy whose quest for the show-business riches of the underworld led him to the starring role in a tragic drama—a macabre cavern carnival that held the whole nation in morbid thrall.

Floyd Collins was a product of his time and place—and in a strong sense he was a victim of the bizarre hostilities that came to be known as the Kentucky Cave Wars.

The Kentucky farm folk who worked the hardscrabble land around America's original speleological frontier knew that Mammoth Cave, while indisputably the largest, was by no means the area's only formidable hole in the ground. Salts Cave, just across the narrow Houchins Valley on Flint Ridge, had been known as long as Mammoth had, and nearby Colossal was discovered in 1895. Martel himself, a 1912 visitor to Mammoth, expressed the belief that the Flint Ridge caves and Mammoth were connected and that the combined system would total 150 miles—a prophecy that proved to be conservative.

Such predictions offered glittering possibilities, but the reality was that Mammoth, still owned by the Croghan family, had a virtual monopoly on the tourist money flowing into the limestone country. Numerous and sundry would-be cave operators were determined to change that situation.

Among them, for example, was Edmund Turner, a young civil engineer who showed up in 1912 and pored over Flint Ridge in search of a new show cave. Three years later his efforts were rewarded by the discovery of a cavern he named Great Onyx Cave. Its glories included a corridor lined with giant columns of onyx and translucent alabaster and, 250 feet below the earth's surface, a huge, statue-like rock formation that was named after the Virgin Mary. Turner's dreams of wealth were short-lived, however; he died of pneumonia less than a year later.

The most successful of the Kentucky prospectors was George Morrison, a Louisville mining engineer who was convinced that Mammoth extended far

The Taming of Carlsbad Cavern

When New Mexico's Carlsbad Cavern was opened to the public in 1911, the owners made few concessions to the comfort—or even the safety—of visitors. The journey into the cave began in a singularly inelegant fashion: Tourists were lowered by guano bucket. After completing the slow, swaying descent, they were handed unwieldy kerosene lanterns to find their way about in the gloom. They quickly discovered that trails were, for the most part, inadequate: Often the tourists had to clamber over fallen rocks and creep along narrow ledges at peril to life and limb.

Improvements were gradually introduced by the National Park Service after it took over the property in 1923. Beginning in 1925, the construction of stairways put an end to the use of the guano bucket and gentled the more treacherous descents within the cave. But climbing back to the surface proved too strenuous for many tourists. By the 1950s the stairs had been replaced by smoothly paved switchback trails, which negotiate the steep slopes in a series of hairpin turns *(below)*.

In 1927 the first electric lights were installed in the cave, and the following year an underground lunchroom began serving food in a large cavern just off the tourist trail, 1.75 miles from the entrance. Employees had to carry food and supplies in from the surface until 1931, when elevators took over the task.

All these amenities have helped turn Carlsbad into one of the nation's great tourist attractions. The cavern's breathtaking limestone formations *(overleaf)* are now viewed by some 800,000 tourists each year.

Zigzagging trails provide an easy means of descent into Carlsbad's Natural Entrance. On summer evenings, visitors gather in the wedge-shaped amphitheater nearby to watch thousands of bats emerge from the cave.

Strategically located near the midpoint of a walking tour of the cave and about 750 feet underground, Carlsbad's modern lunchroom serves as many as 2,000 box lunches an hour.

Three tourists and their guide prepare to explore Carlsbad's largest chamber, the Big Room. The 12-acre subterranean vault houses stalagmites that climb as high as 60 feet.

The aptly named Bashful Elephant, located on the floor of the Green Lake Room, stands only one foot high. Such formations occur when flowstone encrusts an existing rock.

One of Carlsbad Cavern's most delicate formations, the diminutive Dolls' Theater is approximately three feet high, six feet wide and between three and four feet deep. Its tiny stalactites are mere fractions of an inch thick.

The dome-shaped Temple of the Sun rises 30 feet under a canopy of stalactites. Traces of iron oxide have tinged the stalagmite pale yellow.

beyond the boundaries of the Croghan land. If it did, anyone who discovered—or blasted—a new entrance outside the Mammoth tract could go into business for himself. In 1915, Morrison acquired mineral rights allowing him to probe beneath some land adjoining the Mammoth property. He made illicit surveys of the cave itself, sneaking in at night through an unused entrance on the Mammoth property. He then began drilling toward the cave on the land he had leased, explaining disingenuously that he was searching for oil. He sent men into the cave to listen for the drill and to detonate small charges of dynamite while he scanned the terrain above for telltale wisps of smoke. Mammoth officials caught him on their property once and took him to court, where he was fined $75 for trespassing.

After several years of on-and-off efforts, Morrison found what he was after: a back door to Mammoth on his own property. Morrison excavated the opening, advertised it as "The New Entrance to Mammoth Cave," built a hotel and installed a stairway to the wonders below, which turned out to include onyx draperies and waterfall formations more impressive than any in the cave's "historic" section. Even more gratifying from Morrison's standpoint was the fact that his entrance was closer to the main road—and thus more convenient to tourists—than the original one, enabling him to capture a substantial amount of business. Fending off a court challenge by the Mammoth owners, Morrison prospered for several years.

The so-called Cave Wars were precipitated during the 1920s by the bitter rivalry among Colossal, Great Crystal, Great Onyx, Mammoth Onyx, Diamond Cave, Morrison's New Entrance, the venerable Mammoth itself and others. Billboards promised "Kentucky's Most Beautiful Cave" or "The Greatest Cave of All." Agents for the various attractions roamed the major roads, sometimes flagging down motorists and leaping onto their running boards to deliver their spiels. Often they resorted to removing one another's signs or impersonating policemen to intimidate suspecting tourists (because of the visored caps they wore on such occasions, the cave pitchmen came to be called "cappers"). These uniformed impostors sorrowfully advised Mammoth-bound travelers that the tour was too long and difficult for most people, or that they had already missed the turnoff. Shills planted among the Mammoth crowds sneered at the sights and touted the superior grandeur of an attraction down the road. Fights were frequent, and a billboard-bearing truck operated by one promoter so annoyed his competitors that they burned it.

All but lost in the swarm of entrepreneurial hustlers was one man who probably knew more about caves than the rest of them put together. Floyd Collins was a lean, hollow-eyed, uneducated cave-country native whose taciturnity was infrequently relieved by a smile that displayed a gleaming gold front tooth. A member of a large family that tried with marginal success to scratch a living from 200 acres of Flint Ridge farmland, Collins from boyhood had seemed most content while burrowing beneath the earth's surface.

After years of discovering and exploring caves that were either too small or too unimpressive to be commercially profitable, he finally came by accident upon the cavern of his dreams in 1917, when he was 30. Noticing that an animal trap on the family farm had disappeared into a sinkhole, Collins probed around and found a crevice that was "breathing" air. Two weeks of hard digging uncovered a passage leading to a 65-foot-high room garlanded with hundreds of white and cream-colored gypsum flowers. Several other chambers lay beyond. Collins named it Great Crystal Cave, and had it ready for visitors a year later.

Splendid though it was, Great Crystal failed to prosper. It was, alas,

IMPORTANT
NOTICE!

There is prevalent in the region surrounding Mammoth Cave a practice known as "road solicitation." Persons, often in uniform, stop visitors' cars for the purpose of soliciting business for caves other than Mammoth Cave. As a result, those wishing to reach Mammoth Cave may easily be sidetracked from their objective.

The only officially authorized places in this region giving information on Mammoth Cave are the Mammoth Cave Office in Cave City, Kentucky, located at the junction of U. S. 31-W and State Highway 70; and the entrance station of the National Park Service, United States Department of the Interior, at the entrance of Mammoth Cave National Park.

It has become necessary to warn all tourists who wish to reach Mammoth Cave to AVOID INFORMATION GIVEN THEM ON THE ROADSIDE, particularly when cars are stopped for the purpose of giving information to the occupants. The only places in which tickets to Mammoth Cave may be obtained are the Mammoth Cave Office in Cave City, Kentucky; the Frozen Niagara Hotel, located at the Frozen Niagara Entrance to Mammoth Cave; and the Mammoth Cave Hotel, located at the Historic Entrance to Mammoth Cave.

MAMMOTH CAVE OPERATING COMMITTEE

A broadside posted along the route to Mammoth Cave in the 1930s warns tourists that agents for other caves might ply them with misleading information. Mammoth Cave officials estimated that "road solicitation" by rival agents diverted up to a third of their customers.

Smiling solicitors for Great Onyx Cave, one of Mammoth's major competitors, lie in wait for gullible tourists at an information booth along the road to Mammoth Cave. The battle for customers occasionally escalated to the point of burning rival booths.

the last stop (four and a half miles beyond the Mammoth entrance) on the route known as Cave Road; tourists were enticed to rival attractions by the shills long before they reached Great Crystal Cave. Collins concluded that the only way to beat the opposition was to find another cave, with an entrance closer to the main highway than any of the others.

In January of 1925, Collins struck a bargain with three men who owned farms on a patch of high ground a few miles southeast of Mammoth Cave, where the Flint and Mammoth Cave Ridges come together. The landowners agreed to let Collins hunt for a cave on their property and to give him room and board; in exchange they would share among the three of them half of whatever profits might ensue, with Collins to get the other half.

Collins began his search in a hole he had previously spotted beneath a sandstone ledge on Beesley Doyle's farm. He spent three weeks laboriously digging a trench at the entrance and clearing debris from the narrow, twisting and downward-sloping shaft beyond. Never more than a few feet wide, the rubble-strewn crawlway extended diagonally for about 15 feet, dropped straight down for a few feet and then sloped diagonally again, doubling back under itself and narrowing to a vertical drop and a squeeze that Collins could just get through. From there it led diagonally downward again, tapering to a height of only about 10 inches before it reached an alcove barely big enough to turn around in. A few feet farther on lay a narrow, pitlike chute that dropped 10 feet to a shallow cubbyhole and a diagonal, body-sized crevice.

On Friday, January 30, a deadly drama began that was to enrapture the nation. It was the first day Collins was able to get beyond the bottom of the chute and into the crevice, thanks to a dynamite charge he had set off to clear the aperture. Carrying a kerosene lantern and a coiled rope, he wriggled through the diagonal crevice and noticed that its walls and ceiling were composed of dangerously loose dirt and protruding rocks. He emerged on a ledge overlooking a 60-foot-deep pit. Descending on his rope, he investigated the bottom of the pit until his lantern flickered, warning him that it was time to turn back. He climbed back up, leaving the rope in place for his next trip, and crawled headfirst into the body-sized crevice, squirming forward with his hips and shoulders and digging his feet into the floor and walls.

Collins was now 115 feet from the cave's entrance and 55 feet beneath the surface. Before trying to ascend from the crevice into the cubbyhole, he

Mammoth Cave Will Be Closed!

DECEMBER 1st, 1923

Thousands of Tourists Angrily Resent Being Misled as to the Location and Wonders of the Cave. The Mammoth Cave Management Will Find the Remedy.

Mammoth Cave Is A Great Wonder

Mammoth Cave is a Great Wonder because it has what no other cave in the world has, namely: five levels; a navigable river 360 feet below the surface, and four separate routes. Consequently, it attracts to Kentucky tourists from many countries who spend millions of dollars in this State and enrich many of its inhabitants.

The Four Routes

If four separate parties of visitors, each of, say, 50 or more persons, should start from the rotunda near the entrance at about the same time, say 9:00 A. M., and be taken by guides on trips lasting from three to eight hours, these parties could not meet again anywhere in the Cave except at the rotunda where they met for the start.

No One Route Cave A World Wonder

Successful efforts to make of the Mammoth Cave a One Route Cave would reduce it to the level of hundreds of other Caves and destroy its status as what many regard as the greatest of the Wonders of the World and Kentucky's most attractive resort for money spending tourists.

ONLY ONE ENTRANCE

THE ONLY ENTRANCE TO THE MAMMOTH CAVE IS SEVEN HUNDRED (700) YARDS FROM GREEN RIVER. THE FURTHEST BOUNDARY OF THE CAVE IS TWO MILES AND ONE TENTH FROM GREEN RIVER.

So-Called "New Entrance"

THERE IS A SO-CALLED "NEW ENTRANCE" ON A CROSSROAD BETWEEN THE CAVE CITY—MAMMOTH CAVE AND THE GLASGOW JUNCTION—MAMMOTH CAVE ROADS, ABOUT FOUR (4) MILES FROM GREEN RIVER.

The Mammoth Cave Management has nothing to do with this so-called "New Entrance" (said to be about 180 feet deep at the foot of about 260 steps) made by a Development Company, incorporated—not in Kentucky, but at Dover, Delaware, by three residents of that little town, namely:

MRS. L. B. PHILLIPS, A STENOGRAPHER—Three (3) Shares
MISS M. F. VANCE, A STENOGRAPHER—Three (3) Shares
C. H. JARVIS, AN ASSISTANT SECRETARY—Four (4) Shares

That "New Entrance"

To reach any part of the Mammoth Cave from that so-called "New Entrance" a visitor would have to pass:

1.—Through the farm and Cave of the Development Company.
2.—Through the farm and Cave of the Doyel Sisters.
3.—Through the farm and Cave of William Freeman.
4.—Through the farm and Cave of Perry Cox.
5.—Through the farm and Cave of James Hunt.
6.—Through the farm and Cave of W. T. Denison.

Control of the Cave

SINCE 1849 THE MAMMOTH CAVE HAS BEEN UNDER THE SOLE CONTROL OF TRUSTEES APPOINTED BY THE COURT.

IF THERE WERE A NEW ENTRANCE TO MAMMOTH CAVE IT WOULD BE UNDER THE SOLE CONTROL OF THE UNDERSIGNED, LIKE THE REAL AND ONLY ENTRANCE AND EVERY FOOT OF MAMMOTH CAVE.

ALBERT COVINGTON JANIN,
Trustee Mammoth Cave Estate.

Is There Any Onyx In Kentucky?

THE WHITE CAVE

It is situated at about one-half mile from the only entrance to the Mammoth Cave; belongs to the Mammoth Cave Estate and is reached by Automobile or by a fifteen minute walk through very attractive forest scenery.

When the so-called "Great Onyx Cave" was opened to visitors the Mammoth Cave management remembered that our White Cave was described by Dr. Horace Carter Hovey as containing great masses of Onyx. The following is an extract from page 5 of his Mammoth Cave guide book:

"Here is exhibited a splendid piece of Stalactitic drapery called the Frozen Cascade. It is fretted and folded in a thousand fantastic forms and well deserves its name. The resemblance of this mass of Onyx to the gigantic columns formed in winter around great water falls SUCH AS NIAGARA is indeed striking.."

So the name of White Cave was changed to White Onyx Cave since visitors here appeared eager to view formations of the valuable mineral known as Onyx.

Soon thereafter the management was charged in thousands of circulars with deliberately attempting to deceive Mammoth Cave visitors by making them believe that a trip to the "White Onyx Cave" was in reality a trip to the "Great Onyx Cave." This charge was really a libel, but no notice was taken of it by us.

Other Caves then claimed to possess Onyx formation and the question was widely discussed as to what constitutes the valuable mineral known as Onyx and whether it is found in this part of Kentucky.

IN ORDER TO LEARN THE TRUTH OF THE MATTER THE MANAGING TRUSTEE OF THE MAMMOTH CAVE ESTATE ASKED THE GEOLOGIST OF THE STATE OF KENTUCKY WHETHER OR NOT THERE IS ANY ONYX CAVE FORMATION IN ANY OF THE CAVES OF EDMONSON COUNTY, KENTUC-KY.

He answered as follows: (Unimportant parts of the letter omitted.)

"KENTUCKY GEOLOGICAL SURVEY

Frankfort, Ky., October 18, 1921.

I note your query concerning the occurrence of Onyx in Edmonson County, and my answer is as follows:"

"The Onyx referred to in the Great Onyx Cave and other Caves of the Edmonson County region, including Mammoth Cave, is not in fact Onyx at all, but is what is known to Mineralogists as Aragonite. xxx

Real Onyx is fairly hard and is the stuff out of which cameo pins are frequently made. xxx

Aragonite is not hard at all, and you may even scratch it with your finger nail." xxx

Very truly yours,

W. R. JILLSON.

Statement of John M. Hunter

Mammoth Cave Guide

On the morning of Friday, September 7th, 1923, I conducted a party of tourists through Trip No. 1 in the Mammoth Cave. Included in the party were two young men whom I had taken through trip No. 2 in the Mammoth Cave on Thursday night. Mrs. M. A. Bonneford, of Louisville, interfered with me on this trip, while I was endeavoring to supply accurate information concerning Cave formation, by contradicting my statements and by soliciting members of the party to visit the so-called "New Entrance to Mammoth Cave." Mrs. Bonneford spent a great deal of her time in endeavoring to persuade the two young men referred to to visit the so-called "New Entrance," which she told them, according to their statements to me, was "a part of Mammoth Cave." These two visitors to Mammoth Cave came up to me and told me that Mrs. Bonneford told them the "New Entrance" to Mammoth Cave was far superior to "the part of Mammoth Cave" they were in. She said, they told me, "We show the Cave by electric lights and the part of Mammoth Cave shown at the New Entrance is far superior to this part of Mammoth Cave." While we were in the Egyptian Temple I was showing the formations known as the Maiden's Hair and Molasses Candy. Just at this time one of the party asked me if these formations were Onyx, and I said "no, according to what the state Geologist says the formation is not Onyx." Whereupon the tourist said, "Well, what about the Great Onyx Cave? Isn't that Onyx?" I said "The geologist says that it isn't." I also said that the geologists say there is no onyx in any of the Kentucky Caves, and here Mrs. Bonneford interrupted me and said "I know there is onyx. We have black and white onyx both in the New Entrance to Mammoth Cave and we can show it to you."

I said "Well, if there is any onyx anywhere in the cave region it would be in Mammoth Cave. The formations are the finest in the world. If there is any onyx here it would be in Mammoth Cave." She insisted and argued that the cave called "The New Entrance" is a part of Mammoth Cave, and that there is black and white onxy both there. I was much interfered with and annoyed in the discharge of my duty as guide of Mammoth Cave by Mrs. Bonneford, who also took advantage of her trip in Mammoth Cave to solicit patronage for the so-called "New Entrance", giving false information to tourists by telling them that that cave is a part of Mammoth Cave. Any statement that the so-called "New Entrance" is a part of Mammoth Cave is false. No guide of Mammoth Cave has ever shown as a part of any of the four routes for which admission was charged any feature of the cave called "The New Entrance to Mammoth Cave."

JOHN M. HUNTER.

shoved his lantern ahead of him. It fell over—and went out. With his arms lowered to his sides and his feet braced against the floor and walls of the tube, Collins pushed off. His right foot struck a large rock hanging from the roof of the shaft and dislodged it. The rock fell on his left leg at the ankle and pinned it. Collins kicked out with his right leg to try to free himself; he succeeded only in loosening more dirt and stones, which immobilized his right foot as well. He worked his stomach and thigh muscles, his hands still at his sides, and tried to scratch at the dirt and gravel. Every move brought down more debris, binding the helpless Collins tighter in his strait jacket of earth and rocks.

Finally he lay still. He was resting on his left side at a 45-degree angle, his head in the cubbyhole at the bottom of the 10-foot chute. Both arms were pinned at his sides. Both legs were trapped. He was immobilized from the neck down. It was cave-dark. A trickle of water dripped from a limestone boulder onto his cheek. Collins was confronting one of mankind's most terrible and deeply held dreads, the fear of being buried alive.

Alone in his underground prison, shivering in the damp chill, Collins bellowed for help until he lost his voice, then subsided into a fitful doze. He had been trapped at about noon, but he could take some solace in the fact that his partners knew where he was. When Collins failed to return that evening, Beesley Doyle assumed that he had gone to Edward Estes' nearby farm for the night. In the morning, when Doyle checked with Estes and found that Collins had not appeared, both men rushed to the cave with Estes' 17-year-old son, Jewell.

Only Jewell was slender enough to get past the first squeeze in the shaft. Calling Floyd's name as he crawled toward the second squeeze, he heard a faint reply: "Come to me. I'm hung up." But Jewell was afraid to go any farther and retreated.

Word of Floyd Collins' plight spread fast. His 28-year-old brother, Marshall, soon arrived and organized a rescue party from among the two dozen people he found already gathered at the cave entrance, but he too was unable to reach his brother. Homer Collins—at 22 the youngest of the five brothers and, next to Floyd, the most experienced caver—arrived late that afternoon. Stripping to his underwear, he managed to get through both squeezes and down the narrow chute to where Floyd had lain without food, water or light for more than 24 hours. Appalled by Floyd's plight, Homer fed him coffee and sandwiches and began the agonizingly slow work of digging out the dirt and gravel that covered him to the shoulders. The cramped quarters forced Homer to bend and twist into nearly impossible positions to reach the dirt around Floyd. Laboriously filling a can with the debris, he would pass it up to a chain of helpers for disposal and return to digging. Homer toiled through the night and succeeded in uncovering his brother's torso and upper arms. Bone-weary and discouraged, he finally returned to the surface at dawn on Sunday.

While Homer recovered his strength, several men attempted to reach Floyd with food and blankets, but they were defeated by the mud, the pinched crawlways and the oppressive terror that the cave inspired. Floyd was beginning to despair. "I'm trapped, and trapped for life," he told one man who got close enough to talk to him.

Homer went back in again at about 5 o'clock Sunday afternoon. He scooped out more dirt until he uncovered Floyd's hands, still pinned at his sides. He slipped a crowbar into Floyd's left hand so the imprisoned man could try to pry loose the rock that rested on his ankle. Floyd was too weak to move it. Homer meanwhile tried to chip the ceiling rock with a chisel and hammer to give himself more room to work, but this too proved futile.

A desperate 1923 newspaper ad for Mammoth Cave casts doubt on the legitimacy of George Morrison's "New Entrance to Mammoth Cave," denies that there is onyx in Great Onyx Cave, and accuses a competitor of soliciting customers within Mammoth Cave itself.

A few days before he became trapped inside Sand Cave in 1925, Kentuckian Floyd Collins negotiates the narrow crawlway of another cavern. Collins was indefatigable in his search for a new cave that could rival the commercial success of nearby Mammoth.

Homer crawled back out to rest, then returned after midnight as a cold rain pelted the sandstone overhang. Floyd, his body temperature dropping in the cave's 54° F. temperature, was losing his hold on reality. Homer stayed through the night as his tortured brother raved about angels in white chariots and chicken sandwiches and liver and onions. At times, Floyd seemed to recover his senses. "Oh God, Homer," he groaned at one point, "please take me home to bed."

The Louisville papers ran their first stories about the trapped man on Sunday. By Monday, February 2, newspapers across the country were carrying accounts of the rescue efforts, and journalists were beginning to mingle with cave-country residents at the scene. One of the least prepossessing of those who arrived on Monday was the Louisville *Courier-Journal's* William B. "Skeets" Miller, a jockey-sized (117 pounds, five feet five), sandy-haired rookie newsman who looked even younger than his 21 years. Miller immediately approached Homer, who had just emerged from his night-long ordeal at his brother's side and was warming himself at a campfire. "If you want information," Homer snapped, "there's the hole right over there. You can go down and find out for yourself." To his own surprise as well as Homer's, Miller accepted the challenge. "I guess," he said later, "I was ashamed not to."

Putting on a pair of overalls, Miller moved into the opening and squirmed his way downward. After passing through the turnaround he suddenly slid headfirst for several feet and collided with a "wet mass" that groaned and moved. It was Floyd. Fighting a feeling of panic, Miller painstakingly retreated backward up the narrow chute, then slid down again, this time feet first. Each time he became frightened, he noticed, he seemed "to puff up toadlike and become tightly wedged." He lifted the piece of burlap that Homer had placed over his brother's face to keep the dripping water from torturing him. "Put it back," Floyd said weakly. "The water."

Miller started to light a lantern but Floyd said the light hurt his eyes.

Miller began the long, arduous trip back up, thunderstruck by the horror of Floyd's predicament. "As I saw it," he wrote, "nothing would accomplish his release. Reaching him with food would only prolong his agony. Each hour he remained made him less able to help himself. His position was such that those who could get to him only filled the passage above him and could not do much toward rescuing him." Miller's empathy with Floyd transformed him from a mere journalistic witness into an anguished participant in the drama; his articles would win him a Pulitzer Prize.

By Tuesday the Collins story was front-page news in almost every paper in America. Photographers and movie cameramen jockeyed for position amid the growing crowd that swarmed over the muddy clearing next to the shaft—which the press was now calling Sand Cave. The dreadful anguish of a man who could be touched and comforted but not freed seemed to tap some human well of fear and empathy, bringing out both the best and the worst in people.

The rescue attempts were meanwhile bogging down in bickering and acrimony. An attempt to pull Floyd free with a rope attached to a harness around his chest resulted only in afflicting him with excruciating pain. More frantic digging, much of it by Floyd's friend and caving companion Johnnie Gerald, uncovered the trapped man as far as his knees, but still he could not move. The offer of a group of stonecutters to chisel the rock around Floyd and a proposal to drill a shaft to him were both rejected.

Skeets Miller returned to the cave Tuesday afternoon and dug out more dirt until Floyd's knees and calves were exposed. Floyd was lucid and eager to talk. That morning, he said, "I thought, 'Four days down here and no nearer freedom than I was the first day. How will it end? I couldn't think of it. I have faced death before. It doesn't frighten me, but it is so long.' "

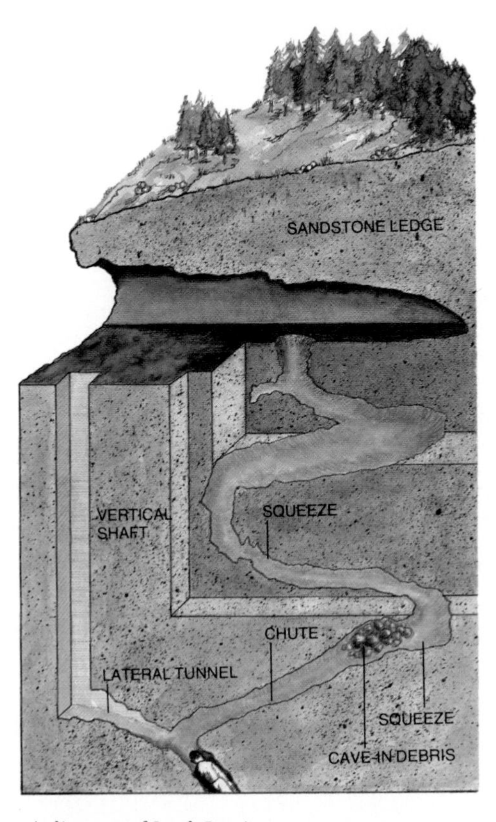

A diagram of Sand Cave's tortuous passages illustrates the difficulties faced by rescuers after Floyd Collins became wedged 55 feet belowground. When their efforts were thwarted by the tiny corridor, they dug a six-foot-wide vertical shaft and a lateral tunnel to reach him.

Since they were now within a few inches of the boulder that pinned him, Miller and Robert Burdon, a Louisville Fire Department lieutenant, decided to try to pry it up with an automobile jack and a crowbar. Miller, descending the shaft yet again, maneuvered the crowbar next to the rock and braced the screw jack against the sloping ceiling, but the jack was too short. Wooden blocks were passed down to fill the space between the crowbar handle and the jack. Then Miller began slowly turning the screw. The crowbar moved. "Keep turning, fella, it's coming off," Floyd yelled. But after a few more turns the blocks slipped out. Miller tried again, but again the blocks moved. "You can do it," Floyd exhorted him. Miller persisted for an hour, adjusting and readjusting the blocks and jack and trying repeatedly to move the rock, but nothing worked. At last his strength gave out. Floyd told him to go get some rest and come back. Sick at heart, Miller prepared to leave, first arranging a burlap-wrapped light bulb on Floyd's chest to give him some warmth.

Shortly after the weary Miller reached the surface at 1 o'clock Wednesday morning, two miners crawled down the passageway and returned with the ominous report that cracks were developing in the ceiling near the top of the chute. Other men sent to investigate came back with even worse news: The ceiling just above the chute leading to Floyd had caved in, making it impossible to reach him. Floyd's light was dimly visible through cracks in the rubble. When Miller arrived back at the cave at about 10 a.m., he asked to see for himself. Accompanied by a miner, he worked his way to the cave-in and called to Floyd. "Come on down, I'm free," came the muffled reply.

Was it possible? Perhaps, Miller thought, the cave-in had actually moved the rock that held Floyd. "Are you sure?" Miller called. Replied Floyd: "Come on down and see."

Hoping that Floyd did not know about the new obstruction and unwilling to tell him, Miller said, "We can't make it right now." Then he remembered that he had left a bottle of milk in a crevice near Floyd's head. He asked Floyd if he could reach the bottle. There was a long pause, then "No, I can't."

Obviously, Floyd's claim had been a desperate enticement for the rescuers to keep on trying. "Then you're not free, are you?" Miller asked. Floyd's reply was almost sullen: "No, I'm not free." Dejected, Miller returned to the surface.

Johnnie Gerald went back in that afternoon, and Floyd overheard him discussing the possibility of shoring up the ceiling and digging out the breakdown debris. "Why don't someone come to me?" he cried. When Gerald told him about the cave-in, Floyd sobbed uncontrollably. At Gerald's suggestion, several men spent the next few hours trying to buttress the ceiling with logs. Gerald worked at the blocked passageway, scraping out dirt and rocks to make a new opening.

At 10:30 that night, Gerald went back in for a final assault. To his horror, he saw that a new cave-in now blocked the shaft above the earlier one. He called to Floyd. "Don't bother me," he heard the delirious man say. "I've gone home to bed and I'm going to sleep." A few minutes later a large rock fell from the ceiling and struck Gerald in the spine. Badly shaken, he turned back toward the entrance. "I'm done," he said to a man with him. "My nerve's gone."

Floyd's position was now more hopeless than ever. He could no longer be supplied even with food and water. Several hours later a 100-pound miner named Roy Hyde led still another attempt to claw a crawlway through the latest breakdown, but it was impossible. "Don't give up," he shouted to Floyd. "We're coming."

"You're too slow," came the pathetic, dispirited words from the man trapped below, "too slow."

On Thursday the character of the rescue effort changed again. Kentucky's Governor William J. Fields, who had previously sent a detachment of National Guardsmen to Sand Cave, dispatched Lieutenant Governor Henry H. Denhardt to take personal command. Denhardt promptly ordered work to

A curious crowd gathers near Sand Cave on the 10th day of Floyd Collins' underground ordeal. Impassioned newspaper accounts of the rescue efforts attracted thousands of visitors to the scene, and additional National Guardsmen were called in to maintain order.

A casket bearing the remains of Floyd Collins lies on display in Crystal Cave, which Collins discovered eight years before his death. "It ought never to have happened," said a local farmer of the tragedy. "Floyd was a caving fool."

begin on the excavation of a vertical shaft. Engineers estimated that the shaft would have to penetrate 55 feet of dirt and limestone to reach Floyd. Because of the danger of further cave-ins, all work would have to be done with pick and shovel instead of explosives. Denhardt also gave orders that no one would be permitted to enter the cave. On Thursday afternoon, teams of volunteers, many of them miners, began to dig a six-foot-square shaft about 20 feet from the entrance to the cave. By Friday they were 10 feet down, but the pace soon slowed as the sides of the shaft constantly spilled dirt into the pit.

As the shovel-wielders inched downward—to 20 feet by Saturday and to 24 feet by noon Sunday—the mass of onlookers, gathered behind a barbed-wire barrier, swelled to festival proportions. Daily front-page stories and frequent radio reports had turned the saga of Floyd Collins into one of the biggest news events of the 1920s. The throng that massed at Sand Cave on Sunday was conservatively estimated at 10,000. "It looked like a country fair," *The New York Times* observed. "Hot-dog vendors, dealers in apples and soda pop, sandwich makers and jugglers vied for the nickels and dimes of the thousands." Floyd's 65-year-old father, Lee, moved through the crowd passing out leaflets touting the wonders of the family's Great Crystal Cave. Tents pitched on the muddy ground housed a field hospital, a kitchen for the workmen and various other support facilities.

The volunteers digging the shaft meanwhile worked doggedly through two days of rain and another day of snow flurries. By Wednesday the 11th, they had reached the 44-foot level, shoring the muddy walls of the shaft to keep them from disintegrating. Two men digging about 50 feet down on Friday the 13th excitedly reported hearing a cough. By that night the shaft was 52 feet deep and, with cave-ins a constant threat, had become almost as dangerous as the cave itself.

The tension was becoming unbearable. Newspapers had been predicting an imminent breakthrough for days. On Saturday the 14th, two weeks after Jewell Estes discovered Floyd's plight, the depth of the shaft reached 55 feet, and the men started digging a lateral tunnel toward the tube where Floyd lay. Shoring as they went along, they worked through Saturday night, Sunday and Sunday night. By noon Monday the tunnel extended just over 12 feet.

The shout from the end of the tunnel at 1:30 p.m. Monday was electrifying: "We're there!" Volunteer Albert Marshall scratched at the opening he had uncovered until it was big enough to let a small man through. Ed Brenner, working beside him, volunteered to squeeze in while the others held his feet. Brenner beamed his flashlight into the dark and saw a man's head six feet below him. He stared for a few seconds, then asked to be pulled out. He uttered a single word: "Dead."

Floyd's body, once again buried to the shoulders, was wedged tightly against the rock over his chest. Water dripped onto his cheek. One eye was partially open. Because the diggers had intersected the shaft above him, they could not remove the body. On the next day, as headlines around the world announced Floyd's death, a coroner's jury descended the shaft to confirm it, and a doctor concluded that Floyd had probably been dead no more than three days. His emaciated condition, the doctor said, suggested that death might have been due as much to starvation as to exposure.

The minister who conducted the funeral service for Floyd Collins eulogized him as a lover of caves "who saw in the gigantic formations and in the fantastic patterns on the walls the traceries of God." Two months afterward a party of miners hired by Homer Collins dug farther down the rescue shaft and removed Floyd's body. They also found the rock that had killed him. Described in the early news accounts as weighing as much as seven tons, it turned out to be a boulder, shaped like a leg of lamb, that weighed 27 pounds.

Lee Collins sold Great Crystal Cave, the cavern that Floyd had discovered on the family farm, to a local dentist in 1927. The new owner exhibited Floyd's corpse in a glass-topped casket in the cave's main hall, a ghoulish bit of showmanship that proved to be profitable. The bitter competition that had sent Floyd Collins to his doom in Sand Cave persisted for years, spreading into such cavern-rich states as Virginia, New York, Tennessee, and especially Missouri. There, a years-long dispute over ownership and control of Onondaga Cave, 80 miles southwest of St. Louis, culminated in the placing of barbed-wire barricades along a 33-foot-wide no man's land through the middle of the beautiful cavern. The dispute was settled in 1935, but deep in the underground reaches of one of the surpassingly lovely caves in America, rusting strands of barbed wire can still be found.

While American caving developed in its unabashedly commercial way, European speleology remained a pursuit for purists. The inaccessibility and danger of many Continental caves discouraged casual visitors, yet these same characteristics posed an almost irresistible challenge to skilled explorers. One member of the elite fraternity of cavers, Norbert Casteret, spoke for them all when he declared that the ultimate sensation was to be caught "in the grip of the demon Adventure, the fascination of the unknown."

Casteret, a native of France's cave-rich Pyrenees region, grew up idolizing Martel as the man "who descended to terrifying depths to look at underground France, and who discovered the palace of the Thousand and One Nights for the greater glory of science." Like Martel, he combined a some-

Ice tinted by minerals forms an emerald drapery in Eisriesenwelt, a rare ice cave nestled high in the Austrian Alps. Frigid bedrock at the cavern's 5,400-foot altitude and air-circulation patterns that chill even warm summer air keep the limestone walls in the depths of the cave sheeted with ice year-round.

what austere and bookish manner with a robust athleticism and an apprecia-tion for the lyrical beauties beneath the earth. As a boy, having been mes-merized by the subterranean fantasies of Jules Verne, he poked around in the numerous caverns in his home province of Haute-Garonne, learning to respect an environment where "every mistake, every act of folly is punished immediately, inevitably and often heavily." He suffered one such painful lesson during his first modest descent by rope when a lighted candle he had fastened to his hatband set fire to his hat.

Casteret's youthful explorations were interrupted by World War I. He enlisted in the French Army in 1915, at the age of 18, and was discharged in 1919. Thereafter he undertook the study of law, a calling that he soon discovered was "not at all adapted to my temperament, which was sportive, venturesome and hardened by war's school of iron and fire." Casteret decid-ed to make caving his livelihood. As he admitted later, the decision "seemed madly ill-considered," but it would turn out to be of enormous consequence for 20th Century speleology.

Throughout Casteret's long career, his motivating force was the explora-tion, not the exploitation, of the wonders that lay underground. In spite of professional attainments that exceeded even his fondest hopes, he would always remain a caving purist. For example, many years after sturdier and more stable craft had been developed, Casteret, "as a sort of tribute to my illustrious predecessor, Martel," insisted on negotiating underworld streams in the same kind of collapsible skiff that Martel had used.

Before he would be able to move ahead in his fledgling profession, Casteret clearly needed to achieve a degree of speleological distinction—and his opportunity arrived in 1922 when, at 25, he was making a system-atic study of Pyrenees caves.

A motionless cascade of ice blankets a slope in Eisriesenwelt, or "world of the ice giants." The icy ramparts, which presented a formidable barrier to early explorers, now attract thousands of tourists to the cavern.

Casteret's searches took him to the hillside village of Montespan, where villagers told him about a body-sized fissure in a neighboring mountain that opened into a water-floored cave. They warned, however, that the cavern ended after 65 yards in a *siphon,* the French word for "sump"—a passageway completely filled by water flowing through it. Squirming through the entrance crack, Casteret stripped off his clothes and waded through a deepening stream until he reached the siphon. There he paused to consider his next step.

His impulse was to force the siphon by diving underwater and feeling his way forward, gambling that he would find air in the passage beyond; if not, that he would be able to turn around and get back safely. Yet to make such an attempt alone in the dark was obviously an act with the potential for immediate and fatal punishment: The siphon might end in a cul de sac; he could plunge down a shaft or emerge in a pocket of gaseous air; quicksand or branches carried by the stream might ensnare him. Against these risks he marshaled the arguments for going ahead: An excellent swimmer, he knew that he could hold his breath for at least two minutes; the looks of the cavern indicated that much more cave lay beyond; there was also what he described as "my habitual obstinacy." He decided to go on.

Placing his lighted candle on a ledge, Casteret plunged in and swam with one hand probing ahead and the other feeling his way along the ceiling.

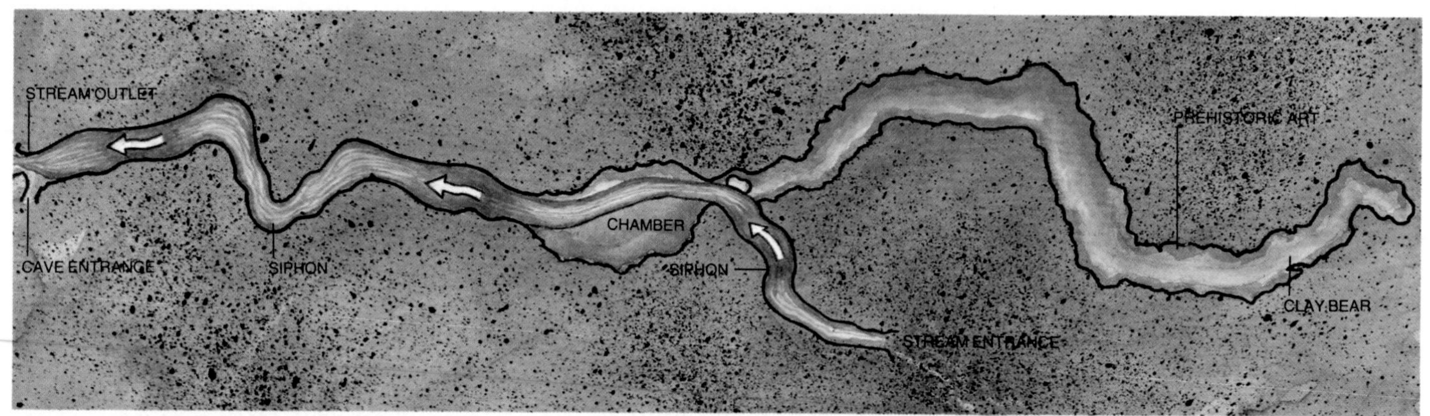

This overhead view of the two-mile-long Grotto of Montespan in the French Pyrenees traces the underground stream that fills bottlenecks, or siphons, in the narrow passageway. French caver Norbert Casteret waded and swam the chill, dark waters to explore the cave in 1923.

After he had progressed a short distance, his head popped to the surface and he could breathe. He turned back immediately, knowing that if he did not he might quickly lose his sense of direction in the blackness. The next day he came back with a supply of candles and matches wrapped up in a rubber bathing cap.

After swimming through the siphon a second time, Casteret lighted a candle in the gallery beyond. By its flickering glow he saw that only a few inches of air separated the water and the pitted limestone ceiling. He paddled on for another 100 yards with his head brushing the rock, then emerged into a room whose ceiling slanted upward to a height of some 40 feet. Beyond this chamber he again waded into the stream—and soon arrived at another siphon. Determined to continue, he steeled himself and dived into this siphon as well. As he struggled to make his way through the stalactites hanging spikelike from the low roof, he began to think for the first time of retreat. This siphon seemed longer than the first; the water was miserably cold. "The loneliness was tremendous," he wrote. "I struggled against an uneasiness slowly turning to anguish."

Still, Casteret persevered. He made it through the second siphon and crawled up the stream floor in a low-roofed gallery, crossed another chamber and walked along a clay bank, leaving footprints he could follow on his return. Gradually losing track of time and distance, he finally reached a bottleneck too narrow to pass through and found a pool filled with branches and tadpoles, a sure sign that an entrance was nearby. The stream disappeared through a tiny aperture to the sunlit world outside, forcing the disappointed Casteret to return the arduous way he had come. In his weariness, he found the longer siphon to be even more of an obstacle on the way back, but he made it through on his second attempt. When he later measured the distance, he found that he had covered nearly two miles in something like five hours.

Casteret made several more excursions into the Montespan grotto before rain swelled the stream and made the underground corridors impassable. The following summer was unusually dry, and he returned—this time with a caving friend, Henri Godin—and explored the maze of passages that extended above and to the sides of the stream course. Casteret and Godin turned into a long side avenue just before the second siphon and continued on until it narrowed to a dirt-floored, low-ceilinged corridor. Here, in a shallow alcove, Casteret found prizes that made all the risks and travail worthwhile. Attacking the clay wall with a pick, he uncovered a chipped flint—proof of human occupation.

Aware that previous cave digs had exposed engravings and crude statues far from the entrances and vestibules where the early cave dwellers lived, Casteret held his candle high and studied the walls. "Suddenly I stopped," he wrote. "Before me was a clay statue of a bear. I was moved as I have

After emerging shirtless from chest-deep waters in the Montespan cave, Norbert Casteret rests by the haunches of a 20,000-year-old clay statue of a bear. Some anthropologists believe that an actual bear's head was originally used to complete the headless figure.

seldom been moved before or since. Here I saw, unchanged by the march of eons, a piece of sculpture since recognized as the oldest statue in the world."

Now, as the two enthralled men looked closer, they saw art all around them—strange symbols etched on the walls with flint, animal drawings, and more statues molded out of clay. Casteret eventually counted some 50 wall engravings and about 30 statues, one depicting a five-foot-long lion marching resolutely toward the entrance. The lion's head and chest were pocked with holes, possibly caused by spears. Archeologists later estimated that the Montespan artists had lived about 20,000 years ago.

Casteret's discoveries at Montespan made him an instant celebrity in the caving world. Maintaining that lofty status, for professional as well as personal reasons, became a lifetime endeavor, with accomplishments measured in statistical records—the deepest, the largest, the highest, the first—or expressed in such abstract superlatives as the "most surprising," the "most picturesque" or the "most enchanting."

One of Casteret's supreme triumphs came in 1926, when he discovered an ice cave that was not only the highest then known but also the "loveliest

A caver paddles through the flooded entrance hall of the Grotto of Cigalère toward a gaping corridor that leads deeper into the cavern. The stream that emerges here has coursed through six miles of the cave's passageways.

and most fantastic." Accompanied by his wife, Elisabeth, his brother and his mother, then in her early fifties, Casteret was crossing a snow field on the Spanish side of the mountainous Pyrenees frontier when he spied an opening in a cliff. The elevation was 8,900 feet. After climbing over a pile of rubble at the entrance, the group gazed down at an ice-walled chamber floored by a frozen lake and bathed in an eerily beautiful blue-green light. The four Casterets moved cautiously across the mirror-like lake and past a translucent column to the edge of an ice-lined shaft. The two men edged around the dangerous pit and examined a frozen waterfall beyond it. Norbert Casteret scrambled up the 30-foot-high cascade with the aid of a pick and wriggled through a small tube to a narrow, high-ceilinged corridor ending in another curtain of ice. By now, the party's candles were running low and their teeth were chattering from the cold. They decided to retreat.

A cold wind sweeping steadily across the icy lake had indicated that there must be a second entrance—and a month later Casteret and his wife returned to find it. Casteret reasoned that the now-frozen river had created the cave during a warmer geologic era thousands of years ago; the stream was, in effect, a fossil. With more time to investigate, he and his wife saw a bird, its wings outstretched in death, frozen beneath 18 inches of floor ice. They skated playfully across this underground rink on their hobnailed boots, a diversion with the therapeutic side effect, as Casteret soberly observed, of warming them up. Reaching the ice curtain where he had stopped before, Casteret managed to ascend it by standing on his wife's shoulders, then pulled her up after him. An upper gallery took them to a round room, where they found a small fissure leading to daylight.

As he squirmed up the final few feet, Casteret was startled to hear a long

Wearing waterproof garb, a caver struggles up a ladder against the icy spray of a waterfall in France's Grotto of Cigalère. Norbert Casteret found the cave during a 1932 survey of potential sources of hydroelectric power in the Pyrenees.

whistle coming from somewhere above him. Popping out of the snow-covered mountainside "like a jack-in-the-box," he found a nuthatch studying him evenly from its rock perch. After reporting the existence and location of the cave to Spanish authorities, the explorer was gratified to learn that his family's discovery would be known as the Grotto Casteret.

That Casteret's unparalleled knowledge of the underground waterways flowing beneath the Pyrenees had practical as well as esthetic value was demonstrated by a project he took upon himself in 1928. His objective was to find the source of the Garonne River, whose tributaries weave through the mountains both above and below ground on either side of the border between France and Spain. A power project proposed by Spain would have diverted a stream that Casteret believed was a major tributary of the Garonne, thus reducing the river's flow in France and raising questions of international law. Systematically tracing the various tributaries, Casteret located the river's ultimate source by using fluorescein dye. His work proved that the river, rising on the south slopes of the Pyrenees in Spain, meanders underground through mountain caves and emerges on the north side in France. This precipitated a complex round of international negotiations, which eventually resulted in the abandonment of the Spanish power project.

In his ceaseless probing of underworld secrets—he discovered and explored more than 2,000 caverns—Norbert Casteret sought satisfaction of the senses rather than practical results or scientific solutions. "You will find in these pages," he once wrote, "neither theories nor technical dissertations on the geology of caves, but simply what I go underground to find and think about; subjects for personal investigation, scenery reminiscent of Dante, new and thrilling sensations."

Yet perhaps more than he realized, Casteret was a scientific generalist: Both to ensure his own survival and to better appreciate what he found in his beloved caverns, Casteret acquired a working knowledge of such varied scientific disciplines as geology and mineralogy, hydrology and hydraulics, paleontology and biology. In so doing he paved the way for a new generation of speleologists already moving underground to apply their own scientific specialties. Ω

A CAVER'S INVENTIVE GEAR

While en route to a French cave in the 1890s, pioneer caver Édouard-Alfred Martel and some companions were halted by a bemused gendarme. As Martel remembered it, the policeman took note of "the extent of our baggage," and asked "if we were a traveling circus, and if we had a license." Martel tolerated such gibes cheerfully because he knew full well that his life depended upon his profusion of equipment.

Modern speleologists have improved many of Martel's methods and devices, and have developed a remarkable array of caving hardware, but his belief in thorough preparation and reliable gear is still the caver's unofficial creed.

Heavy-duty clothes, including high boots, kneepads and a sturdy helmet, prevent abrasion and provide warmth in the chilly underground temperatures—typically near 50° F. in the United States. The modern caver's pack is small enough to fit through tight spots, yet roomy enough to hold such basic supplies as high-energy food, a first-aid kit, a superinsulating "space blanket," water and tools for repair of the carbide helmet light. Experienced cavers carry two reserve light sources: a flashlight and a waterproof packet of matches and candles. A compass, note pad and pencil are used to record the route taken through the mazelike passageways.

Harnesses, strong nylon ropes and other mountaineering gear are essential for cave exploration. A compact steel-cable ladder is ideal for short climbs, but mechanical ascenders are used for longer, rope ascents. A rappel rack distributes rope stress during a vertical descent, permitting a controlled drop, while a spring-gate carabiner ensures reliable harness connections.

Martel, who was the first to use such mountain-climbing gear in a cave, found this equipment so useful, and the ups and downs of caving so demanding, that he issued a provocative challenge to the more flamboyant alpinists of his day: "Try mountaineering reversed for once."

Hooking one leg around a steel-cable ladder so that his heel lodges snugly on the rung, a well-equipped caver begins a journey to the underground. Some basic items of his equipment, from a simple pack to exotic mountaineering gear, are detailed at right.

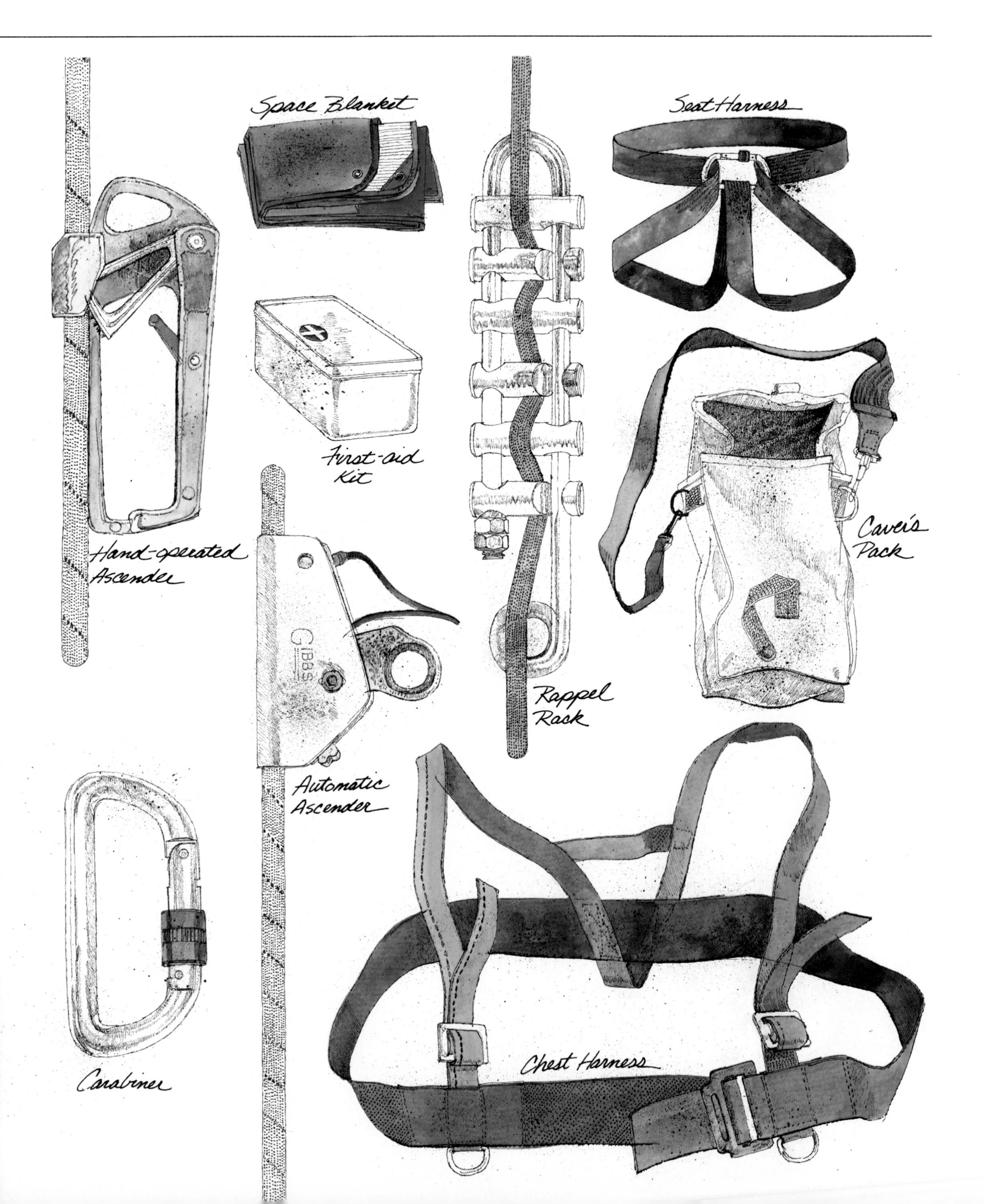
Space Blanket
Seat Harness
First-aid Kit
Hand-operated Ascender
Caver's Pack
Automatic Ascender
Rappel Rack
Carabiner
Chest Harness

A Breathless Squeeze and a Bracing Ascent

Cavers on a subterranean trek may walk for hours as if on a Sunday stroll, but sooner or later they will have to bend, crawl, squeeze or wriggle through a tight spot. Sometimes it takes a cool head as well as agility to get through.

Diminishing headroom may force a caver to adopt a crouchlike crawl, with the weight on the hands and toes rather than on the knees. When crawling on hands and knees is required, kneepads protect against painful and potentially crippling kneecap injuries. And if the cave ceiling continues to drop, a head-first belly crawl may be in order.

Cavers quickly learn that they can squeeze through a surprisingly small opening: An adult can usually make it through an aperture one foot wide. And while getting stuck is the ever-present caving nightmare, experience teaches that the seemingly viselike grip of the rock can be escaped by patient movements and exhaling to reduce chest size.

At times, the proximity of cave walls may actually assist the explorer. By bracing the back and legs against opposite walls, a caver can move vertically or horizontally above streams and obstacles. Called chimneying, this technique is used for short passages, and only when it is certain that the walls do not diverge beyond the span of the caver's legs.

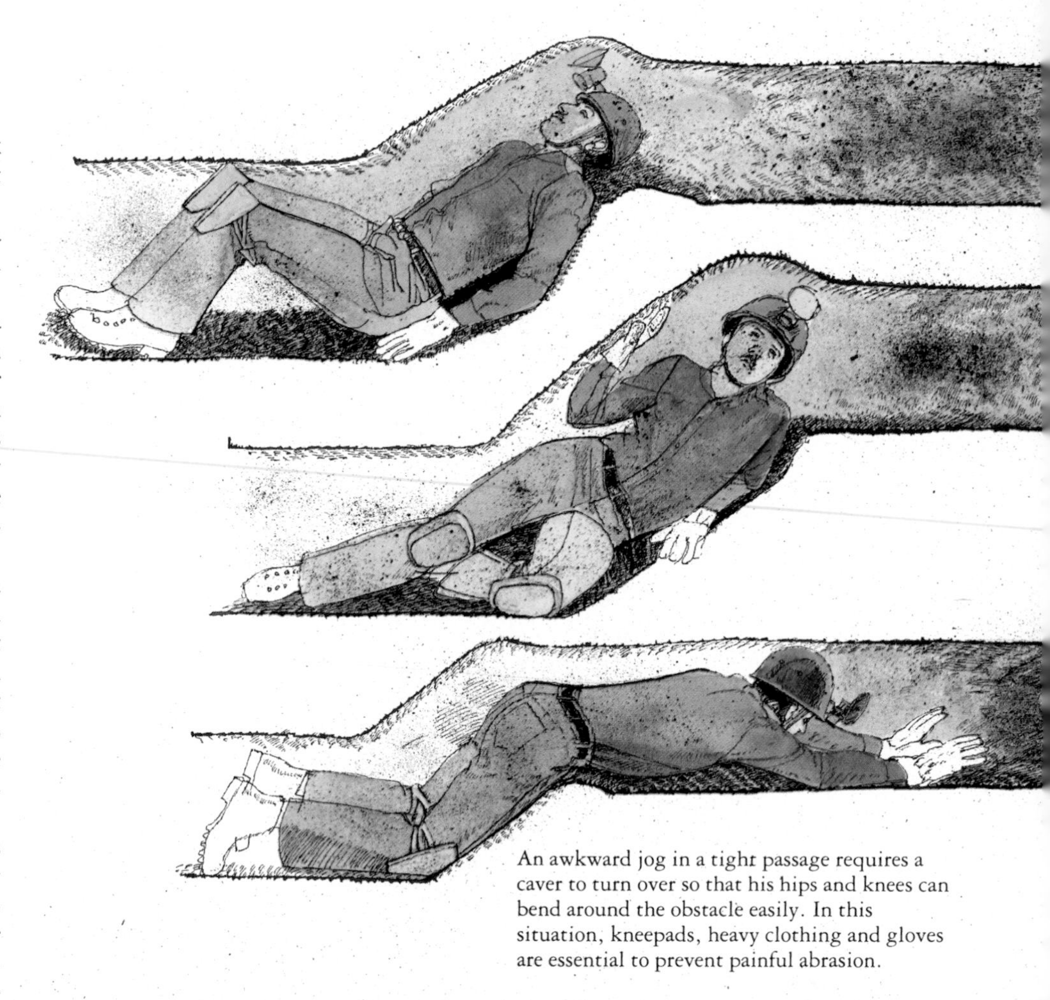

An awkward jog in a tight passage requires a caver to turn over so that his hips and knees can bend around the obstacle easily. In this situation, kneepads, heavy clothing and gloves are essential to prevent painful abrasion.

His helmet removed to clear a low ceiling, a caver slithers through a cramped corridor. All extra gear has been discarded, and his pockets have been emptied to minimize impediments.

Chimneying up a narrow vertical shaft, a climber *(left)* braces his feet and back on opposite walls and squirms upward with small steps. Another method *(right)* involves placing one foot on each wall for greater stability while the arms and hands help support the back.

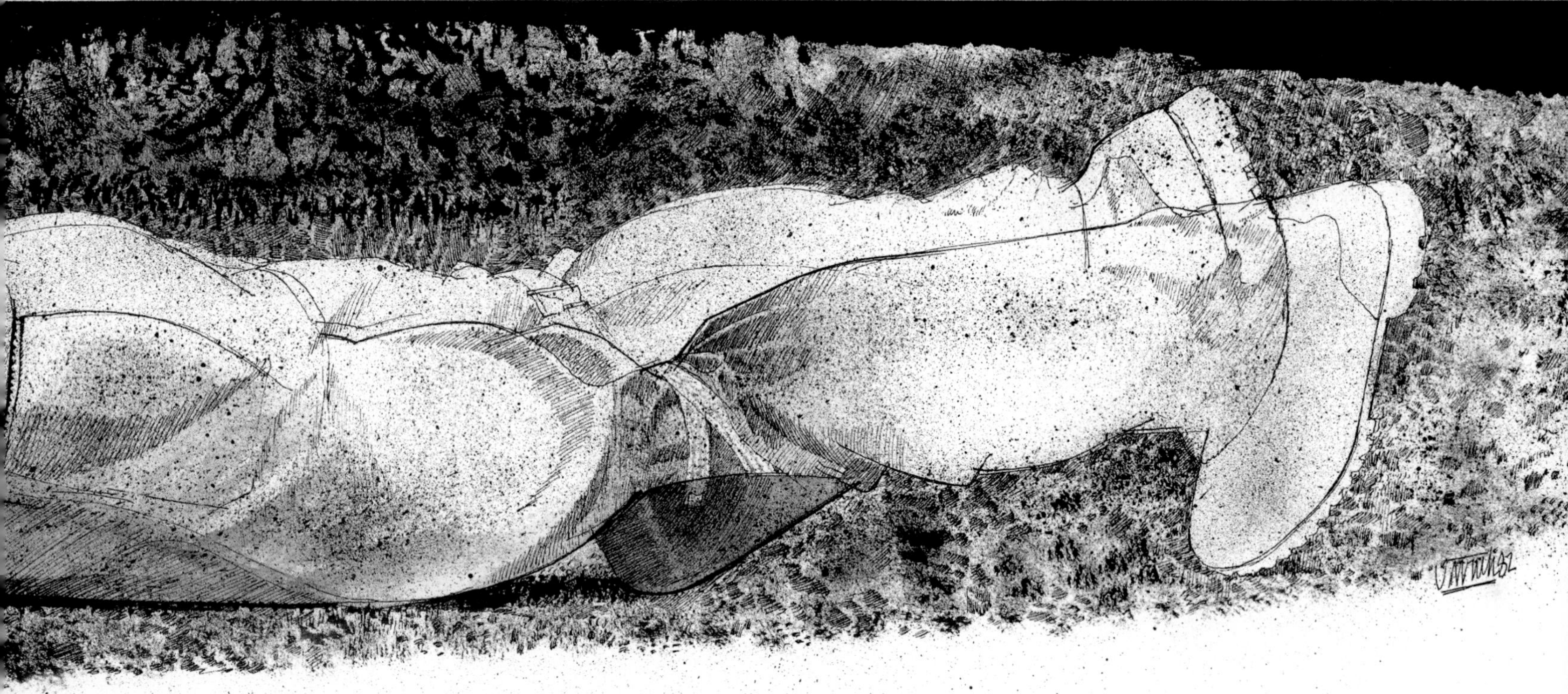

A caver's hands and feet provide four contact points to control his descent in a chute. Most of his braking is done by pressing the broad surface of his back against the wall.

His feet planted firmly on opposite walls, a caver employs a technique known as canyoning *(left)* to move forward horizontally over a narrow chasm. Even when a descent is relatively easy *(right),* it is good practice to always maintain four points of contact by using handholds.

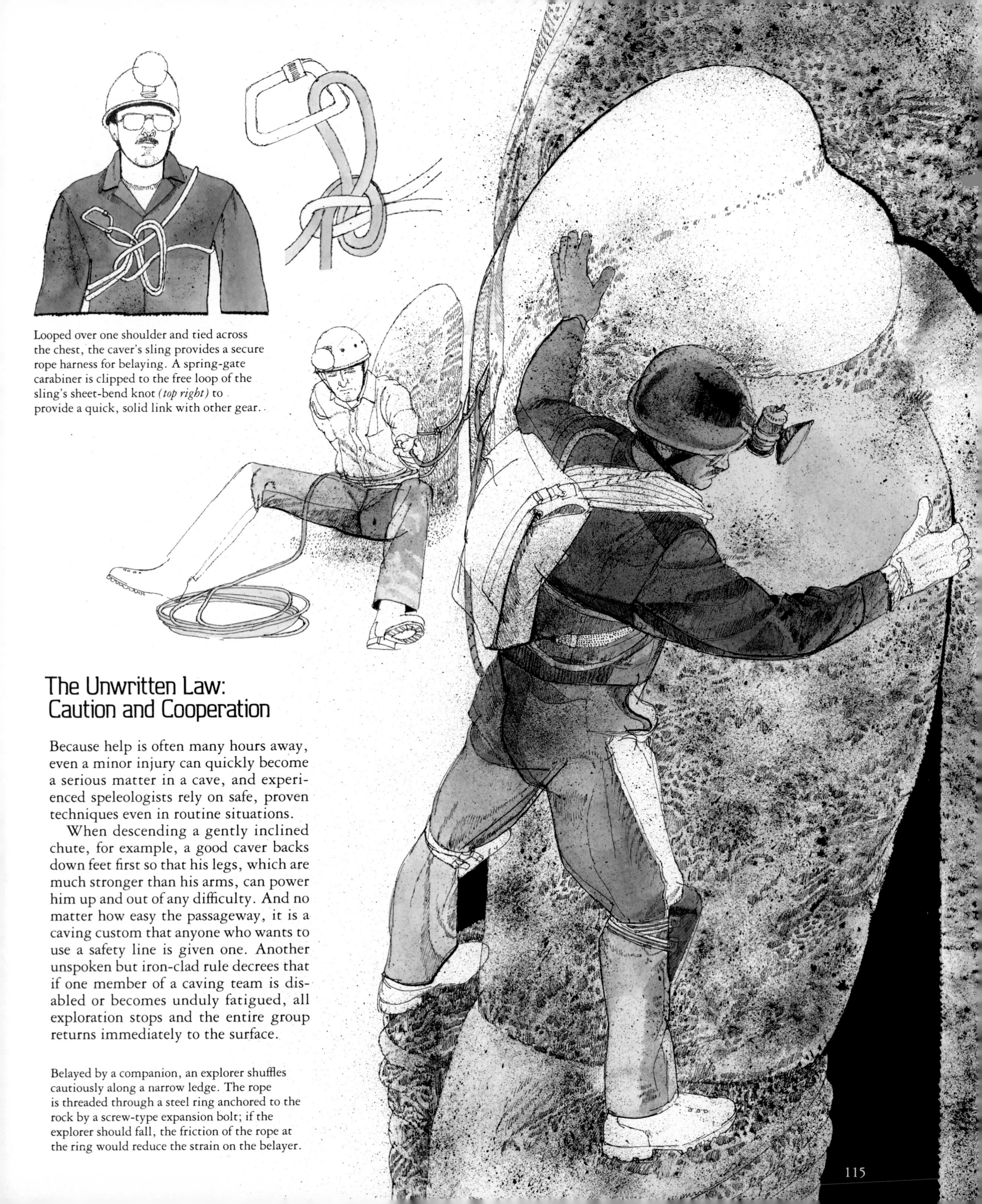

Looped over one shoulder and tied across the chest, the caver's sling provides a secure rope harness for belaying. A spring-gate carabiner is clipped to the free loop of the sling's sheet-bend knot *(top right)* to provide a quick, solid link with other gear.

The Unwritten Law: Caution and Cooperation

Because help is often many hours away, even a minor injury can quickly become a serious matter in a cave, and experienced speleologists rely on safe, proven techniques even in routine situations.

When descending a gently inclined chute, for example, a good caver backs down feet first so that his legs, which are much stronger than his arms, can power him up and out of any difficulty. And no matter how easy the passageway, it is a caving custom that anyone who wants to use a safety line is given one. Another unspoken but iron-clad rule decrees that if one member of a caving team is disabled or becomes unduly fatigued, all exploration stops and the entire group returns immediately to the surface.

Belayed by a companion, an explorer shuffles cautiously along a narrow ledge. The rope is threaded through a steel ring anchored to the rock by a screw-type expansion bolt; if the explorer should fall, the friction of the rope at the ring would reduce the strain on the belayer.

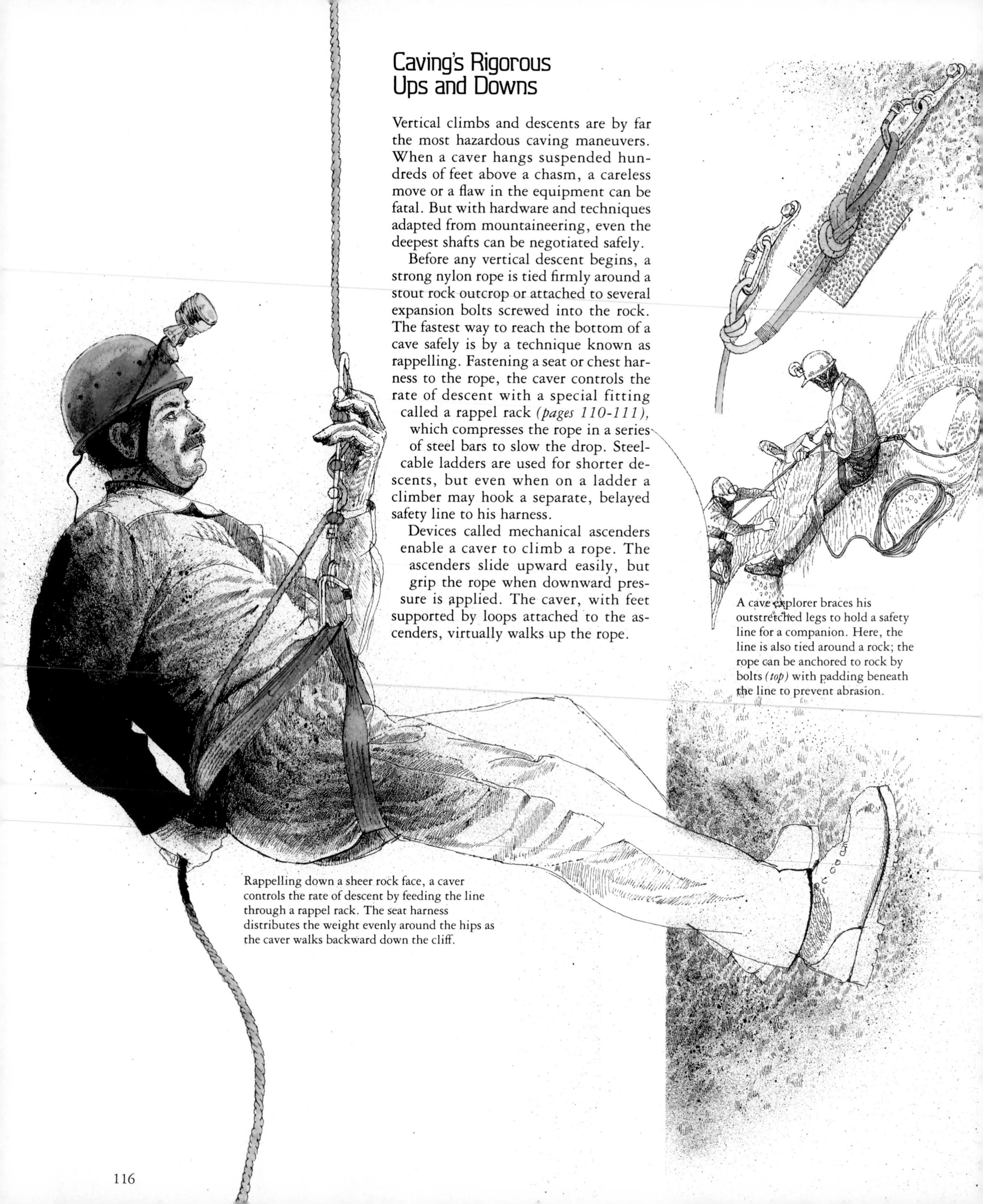

Caving's Rigorous Ups and Downs

Vertical climbs and descents are by far the most hazardous caving maneuvers. When a caver hangs suspended hundreds of feet above a chasm, a careless move or a flaw in the equipment can be fatal. But with hardware and techniques adapted from mountaineering, even the deepest shafts can be negotiated safely.

Before any vertical descent begins, a strong nylon rope is tied firmly around a stout rock outcrop or attached to several expansion bolts screwed into the rock. The fastest way to reach the bottom of a cave safely is by a technique known as rappelling. Fastening a seat or chest harness to the rope, the caver controls the rate of descent with a special fitting called a rappel rack *(pages 110-111)*, which compresses the rope in a series of steel bars to slow the drop. Steel-cable ladders are used for shorter descents, but even when on a ladder a climber may hook a separate, belayed safety line to his harness.

Devices called mechanical ascenders enable a caver to climb a rope. The ascenders slide upward easily, but grip the rope when downward pressure is applied. The caver, with feet supported by loops attached to the ascenders, virtually walks up the rope.

A cave explorer braces his outstretched legs to hold a safety line for a companion. Here, the line is also tied around a rock; the rope can be anchored to rock by bolts *(top)* with padding beneath the line to prevent abrasion.

Rappelling down a sheer rock face, a caver controls the rate of descent by feeding the line through a rappel rack. The seat harness distributes the weight evenly around the hips as the caver walks backward down the cliff.

Using mechanical ascenders, a caver steps up a rope, pulling the ascenders along by hand. A chest harness attached to the rope provides balance for a brief rest. When the caver places his weight on the ascender, a toothed jamming arm grips the rope.

Another type of mechanical ascender automatically follows upward movement and does not have to be raised by hand. In this arrangement, one of several used by cavers, an ascender is attached to one foot, a middle ascender is raised by an elastic cord attached to a chest harness, and a third ascender, secured at shoulder level, keeps the climber upright.

germinate in the guano and, like potatoes in a cellar, sprout without benefit of light into stunted white plants—"blanched stalks," as Humboldt described them, "with some half-formed leaves." Even if sunlight were available to permit further growth, the plants would not last long; beetles, four-inch-long cockroaches and cave crickets would flock to the ghostly little forest and devour the precious vegetable matter.

The twilight zone of the cave, teeming with visitors and littered with the nutritious by-products of their activities, offers a cornucopia to its trogloxene residents, compared with the dark and barren inner reaches that constitute the domain of the troglobites. Here live creatures whose links to the ordinary world of sunshine, rain and green growth have been ruthlessly obliterated by the operation of natural selection during eons of time. No habitat on earth presents more dramatic examples of the cardinal principle of evolution: The traits that jeopardize the survival of a species gradually degenerate, while the characteristics that enhance continued existence develop and strengthen.

The survival of the troglobites, in their black world of eternal famine, depends above all on success in their search for food; nearly all their sensory and physical resources are devoted to that quest, and any organ unnecessary to it has degenerated or perhaps disappeared entirely. Thus, without exception, all true troglobites, living as they do in total darkness, are blind. Since they need neither protection from the sun's burning rays nor coloration to camouflage them from the eyes of predators, the troglobites lack pigment and often appear a sickly white.

While the permanent cave dwellers have lost attributes such as sight and pigmentation, they have developed new characteristics to help them make the best of a bad deal. Many, for example, are assisted in their hunt for food by antennae that are longer and more sensitive than those of their surface cousins. Since food is required to provide energy, and the cave larder is scanty at best, the troglobites are masters of energy conservation: They generally have longer legs than their outside counterparts do, and they can range farther with less strain on their metabolisms. Where most animals tend to be food specialists, troglobites cannot afford to be as picky; nearly all of them can and do eat anything that might provide even the slightest bit of nourishment.

Despite a general dependence on their more peripatetic cave companions to provide them with gleanings of sustenance, there are certain periods when troglobites dwell in a realm of relative plenty. Heavy rains and snow melts bring floods, and the rushing waters carry into the caves a splendid variety of foods—leafy twigs and branches, logs infested with worms or insects, even the carcasses of drowned animals. Although there are no seasonal changes in the constant temperature zones of the caves and no apparent signals of the passing of time, certain cave denizens somehow sense the coming spring floods—and prepare for their own regeneration.

In Shiloh Cave in Indiana, for example, crayfish develop sperm and eggs in late summer or early autumn. Mating probably occurs then, but the female stores the male's sperm in a special receptacle. When the spring floods arrive, she fertilizes the eggs in the process of laying them, and they hatch in midsummer—a period when the young can feed on peak populations of such tiny crustaceans as copepods, isopods and amphipods. These lesser crustaceans have themselves been nourished by bacteria and protozoans that fed on organic matter brought in by the floods.

It is widely believed that the animals living in caves were, to one degree or another, preadapted; that is, even in their precave generations they had physical, sensory or behavioral characteristics particularly suiting them to life underground. The blind salamanders that crawl lazily through the waters of widely separated European and American caves admirably support this idea: Whether on the surface or below, salamanders are endowed with low metabolisms; they require moisture to breathe, making caves, with their high humidity, an especially friendly habitat; and even surface salamanders prefer dark places beneath rocks or in mud and slime. But it is in the differing evolutionary stages of salamanders found in caves far removed from one another that the reclusive animals are of most interest to biospeleologists.

The first scientific mention of a cave-dwelling salamander appeared in a book published in 1689 by the Slovenian Baron Johann Valvasor. A caving enthusiast when he was not bearing the burdens of title, the baron spotted the creature in a stream in a Yugoslavian cave. Pinkish white, eyeless, perhaps a foot long, supported by an odd set of front feet with three toes and by hind feet bearing two toes each, this unlikely beast looked as though it could be a new and pint-sized species of dragon. Local villagers had long believed that a dragon living in a cave at the source of the River Bella caused periodic floods by opening sluice gates when its living quarters were threatened by rising water.

The truth, as it gradually emerged over the next two centuries, was hardly less fantastic. The miniature dragon was actually a salamander of the type that came to be known as *Proteus,* the name of a sea-god capable of changing its form. Born with rudimentary but sightless eyes, *Proteus* loses them as it matures, until not even the sockets are recognizable. It changes from a dark gray color at birth to a pigmentless white, tinged slightly by its blood vessels. It possesses both functional lungs and a set of feathery red gills on each side of its neck. *Proteus* appears to be sensitive to light in spite of its blindness, although precisely how remains one of its many secrets.

So strange an animal was worthy of even royal attention: Archduke Jean of Austria kept a specimen in a grotto at his country house for eight years. Formal scientific recognition and a dignified name *(Proteus anguinus)*

White fungus flourishes on a thick, decaying layer of nitrogen-rich bat guano that blankets the floor of a cave in Missouri's Ozark Mountains. Fungi are among the few forms of living vegetation that contribute to the sparse food chain in caves.

A tiny millepede, eyeless and devoid of pigment after eons of adaptation to the perpetual darkness of a cave, somehow senses a beam of light and curls up in a frantic effort to avoid it.

stripped *Proteus* of its reputation for having dragon powers, but made it even more interesting to scholars, many of whom obtained one for their laboratories. These lab specimens astonished scientists by demonstrating their ability to survive for extended periods of time without apparent nourishment: One collector avowed that his *Proteus* succumbed to starvation only after 14 foodless years. Modern students of *Proteus* are skeptical of such claims, but suggest that a mature specimen in a cool environment could survive as long as three years without any food whatever.

Later investigators added more pieces to the still-incomplete portrait of this bizarre creature. During the mating season in May—by no coincidence a month with a high flood rate—the male *Proteus* establishes a territory and challenges any trespassing male, flailing away with its tail and biting the intruder if it does not leave; how *Proteus* senses the alien presence is unclear. It was once believed that in cool temperatures *Proteus* bore live young like a mammal, but that it laid eggs like a reptile if the water was warmer than 59° F. In 1958, however, biologists were finally able to prove that the females always lay eggs, usually under large underwater rocks, regardless of the temperature. Unlike its relatives outside caves, *Proteus* does not metamorphose from a larva to an adult form, but retains a larva-like form throughout its long life of up to 25 years. Its senses of touch and smell are apparently keen. No one has yet fully explained exactly how or why its eyes deteriorate beneath the skin as it matures; eye development seems to stop just before *Proteus* hatches and to regress afterward.

The first clue that the European *Proteus* had an American cousin came in an 1885 *Scientific American* article that included "white, blind lizards" among the fauna found in Missouri's Marble Cave. Six years later, a single specimen of what would come to be known as the Ozark blind salamander was plucked from another Missouri cave and dispatched to the Smithsonian Institution. Others were subsequently found, and scientists learned that the Ozark blind salamander differs from *Proteus* in several striking particulars. The Ozark salamander, for example, makes occasional forays to the outside world, apparently because it lacks the ability of *Proteus* to survive for long periods on little or no food. Even more significantly, the Ozark species metamorphoses from a fairly dark, relatively stocky creature with small but functioning eyes to a slender, pale and sightless adult: Its eyelids grow together as it matures.

As it happens, the process of metamorphosis requires a lavish expenditure of energy; it is therefore a considerable disadvantage in the world of caves, where the food needed to generate energy is in such short supply. In that major respect, *Proteus* is better adapted to cavern conditions than the Ozark blind salamander. If, as scientists assume, the length of time an animal species has spent in the caves can be gauged by the number and degree of cave-adapted features, it seems likely that the ancestors of *Proteus* were veterans of the underground when the forefathers of the Ozark blind salamander were still hiding under rocks on the surface.

Another American member of the family was discovered in 1895, when a crew of well drillers near San Marcos, Texas, saw several skinny white salamanders, of a sort previously unknown, rise to the surface from a stream they had tapped 188 feet underground. Only a few more of the creatures were subsequently seen—until 1938, when the ubiquitous team of Mohr and Dearolf set out to look for them in Ezell's Cave, near San Marcos, where several had reportedly been sighted.

The two men lowered themselves 40 feet into the entrance chamber by rope, then inched their way through a 100-foot-long network of narrow

In this 1678 engraving, a swordsman attacks a cave-dwelling dragon popularly blamed for floods in Lucerne, Switzerland. *Proteus anguinus (left),* the 10-inch-long blind salamander discovered at about the same time, was thought to be the offspring of a similar beast.

crawlways until they reached a pool of motionless water. Stationing themselves beside the pool, Mohr and Dearolf trained their lights on the water and waited. At length they captured an inch-long flatworm—which proved to be a new species—but sighted no salamanders. The second day yielded a tiny white isopod, and a third trip was rewarded with the discovery of a transparent shrimp, visible only because of the shadow it cast on the floor of the shallow pool.

Soon after descending to the pool for a fourth time, Mohr spied another flatworm and waded into the water to snare it in a jar. As he turned back toward shore, his light revealed a strange white form about four inches long, idling in half a foot of water. Breathless with excitement, Mohr kept his eyes on the salamander—it was "beautiful, exotic, with a silky white body and brilliant, tufted blood-red gills"—while Dearolf hastened to help with a net. Afraid to wait, Mohr gingerly maneuvered an open jar toward his quarry, but the salamander swam rapidly away. Mohr made another pass with his jar and missed. The salamander was gone.

Suddenly it reappeared nearby. "I scooped again with my hands and missed," he recalled. "Again, and the weird beast was flopping in my hands and up onto my close-pressed wrists. I was begging Dearolf to hurry and in another moment he was there with the net and a vacuum bottle." Triumphant, they examined their prize. The Texas species had even more degenerate eyes than *Proteus,* with sensitive vibration-detecting nerves on a tapering head and a blunt spoonlike snout, and long toothpick-like legs that ended in a cluster of appendages bearing a startling resemblance to tiny human hands. In each of these aspects the Texas salamander was better equipped than *Proteus* for cavern exigencies, and, if evolutionary theory is correct, the forebears of the New World variety may well have been isolated in caves even before those of the Old World champion.

The discoveries of the Ozark and Texas blind salamanders have been followed in recent years by the documentation of seven more species in America—four in Texas, one in Georgia and two in the Appalachian region. They differ from each other in bone structure, pigment, length of limbs, degree of eye degeneration and many other features—all milestones of adaptation.

The creatures that creep across cave floors, skitter along the walls or weave soundlessly through the streams—the spiders, crickets, beetles, millepedes, springtails, shrimp, amphipods, isopods and dozens of others—

bespeak the endless ingenuity of evolution. There are, for instance, more than 200 known species and subspecies of troglobitic beetles in American caverns alone. One type of cave cricket, equipped with antennae four times as long as its body, has become so adapted to the unchanging temperature belowground that it cannot survive in the outside world.

The blind rhadinid beetle, another well-adjusted troglobite, dines almost exclusively on the eggs that crickets deposit in the silt of the cave floor. One of the few cave species with such a specialized diet, the rhadinids are highly adapted for it; sensory organs on their antennae guide them to the eggs, which they puncture with their sharp tonglike mandibles. Another cave-dwelling beetle ensures the survival of its kind by laying fewer and larger eggs than its relatives aboveground. The yolks contain so much nourishment that the young need no food at all for the seven to nine months it takes them to mature into adults.

The glowworms of the Waitomo Caves in New Zealand are the only cave creatures that are equipped with their own light source. The glowworms, actually the larvae of a species of gnat, cluster on the dark ceilings and walls, creating a galaxy of blue green lights. Beneath each shining bug is a sticky thread known as a "fishing line," as much as a yard long, that snares flying insects attracted by the light. The glowworm reels

When the three-to-four-inch-long Ozark blind salamander is nearing the end of its larval stage *(top)*, it has functional eyes and gills; in its adult stage *(bottom)*, the eyes have degenerated and the gills have disappeared.

in its victim, consumes it, and then tidies up the fishing line before deploying it again.

Even after decades of study, the teeming world of the cave creatures remains capable of presenting biospeleologists with a complete surprise. A film shown at the Eighth International Congress of Speleology in Bowling Green, Kentucky, in 1981 documented the recent discovery of one of the most singular cave creatures ever found, a crustacean that is a member of not just a new species, but also a new genus, a new family, a new order and a new class. Four specimens of this blind, unpigmented animal were found by biospeleologist Jill Yager and her cave-diving companions in an underwater cavern on Grand Bahama Island.

"It looked like a worm with legs," said a colleague, John R. Holsinger, who was summoned to inspect the find; "It's apparently an evolutionary relict." Holsinger explained that most crustaceans have specialized body segments with different appendages used for different purposes. The newly named *Speleonectes lucayensis,* by contrast, has a series of undifferentiated and unspecialized segments. It is apparently a direct and little-changed descendant of a primitive crustacean that swam the tropical seas millions of years ago and evolved into hundreds of more highly developed forms. *Speleonectes* is, in effect, a living fossil.

The cave habitat, with its unique stability and simple food chain, is as close to a self-contained ecosystem as any in nature. It is conceivable that the cave animals could survive the various horrors that threaten life on the surface—toxic fumes, disruption of the food chain, even nuclear war. The lessons derived from the study of this ecosystem and the creatures that have made themselves over in order to thrive in it, according to Holsinger, offer instructive examples to humans, who need to learn about "population control or perhaps energy conservation." More generally, Holsinger observes, "We're going to have to evolve ways to get along with less"—a dilemma solved long ago by the sturdy masters of making do underground. Ω

This many-legged creature, a previously unknown class of crustacean, was first seen in 1978 swimming at a depth of 60 feet in the total darkness of Lucayan Cavern in the Bahamas. The ancestors of the inch-long *Speleonectes* (cave swimmer) may have been washed into the cavern from the sea millions of years ago.

Glowworms, larvae of a cave gnat, create a celestial impression on the ceiling of a grotto in New Zealand's Waitomo Caves. The luminescence lures flying insects to sticky fibers deployed by the glowworms to trap their prey.

THE HARDY DENIZENS OF DARKNESS

"Natural selection is daily and hourly scrutinizing, throughout the world, every variation, even the slightest; rejecting that which is bad, preserving and adding up all that is good; silently and insensibly working, whenever and wherever opportunity offers, at the improvement of each organic being in relation to conditions of life."

This observation by Charles Darwin, published in 1859 in *The Origin of Species,* applies nowhere more accurately than to the innermost reaches of caves and to the troglobites, the subterranean fauna that inhabit them. No creatures on earth are better showpieces of his theories or offer more dramatic demonstration of the basic Darwinian tenet that only the fittest survive.

All troglobitic creatures—including certain species of fish, insects, crustaceans and amphibians—evolved from surface-dwelling ancestors whose environments resembled, in some respects of humidity or darkness, that of caves. Long ago, some of these creatures began to colonize caves. After ages of genetic isolation and adaptive change, their progeny became permanent residents.

The successful creatures in this realm tended to be smaller, with slower metabolisms and longer life spans. Their eyes degenerated, since the perpetual darkness yielded no advantage to those who had sight. With no need for protection from the sun or camouflage from enemies, the cave creatures lost their pigmentation. They developed elongated limbs and fins capable of more efficient movement, and highly sensitive sensory organs to detect the presence of predator or prey.

The ghostly Texas blind salamander *(right)* and the other troglobites shown on the following pages epitomize the legions of creatures that—because of the process Darwin so eloquently described—can never return to the surface of the earth, but are wonderfully equipped to survive below.

Breathing through feathery pink gills, a four-inch-long Texas blind salamander turns its spoonlike snout in search of prey, which it finds with highly developed sensory organs along its sides and tapering head.

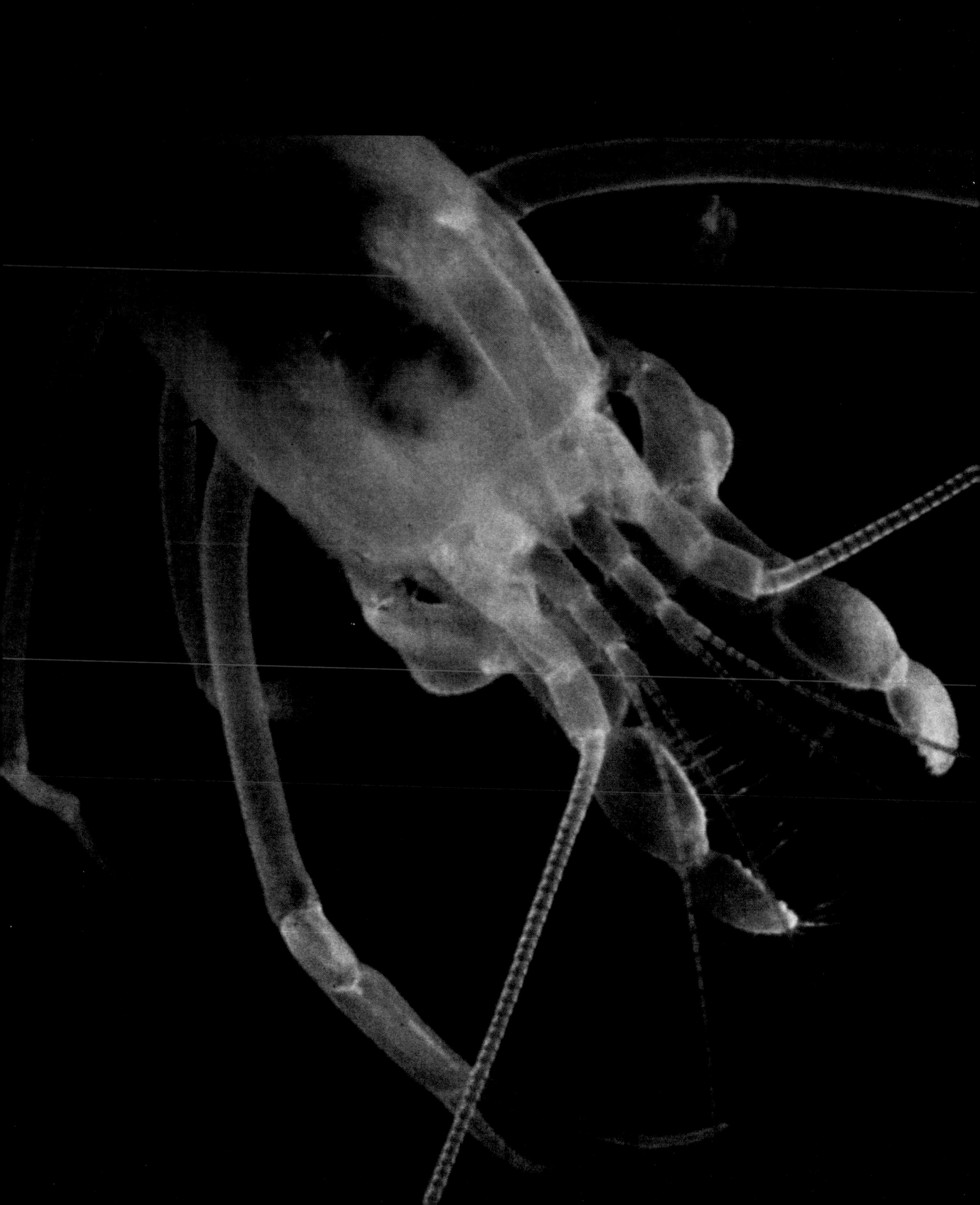

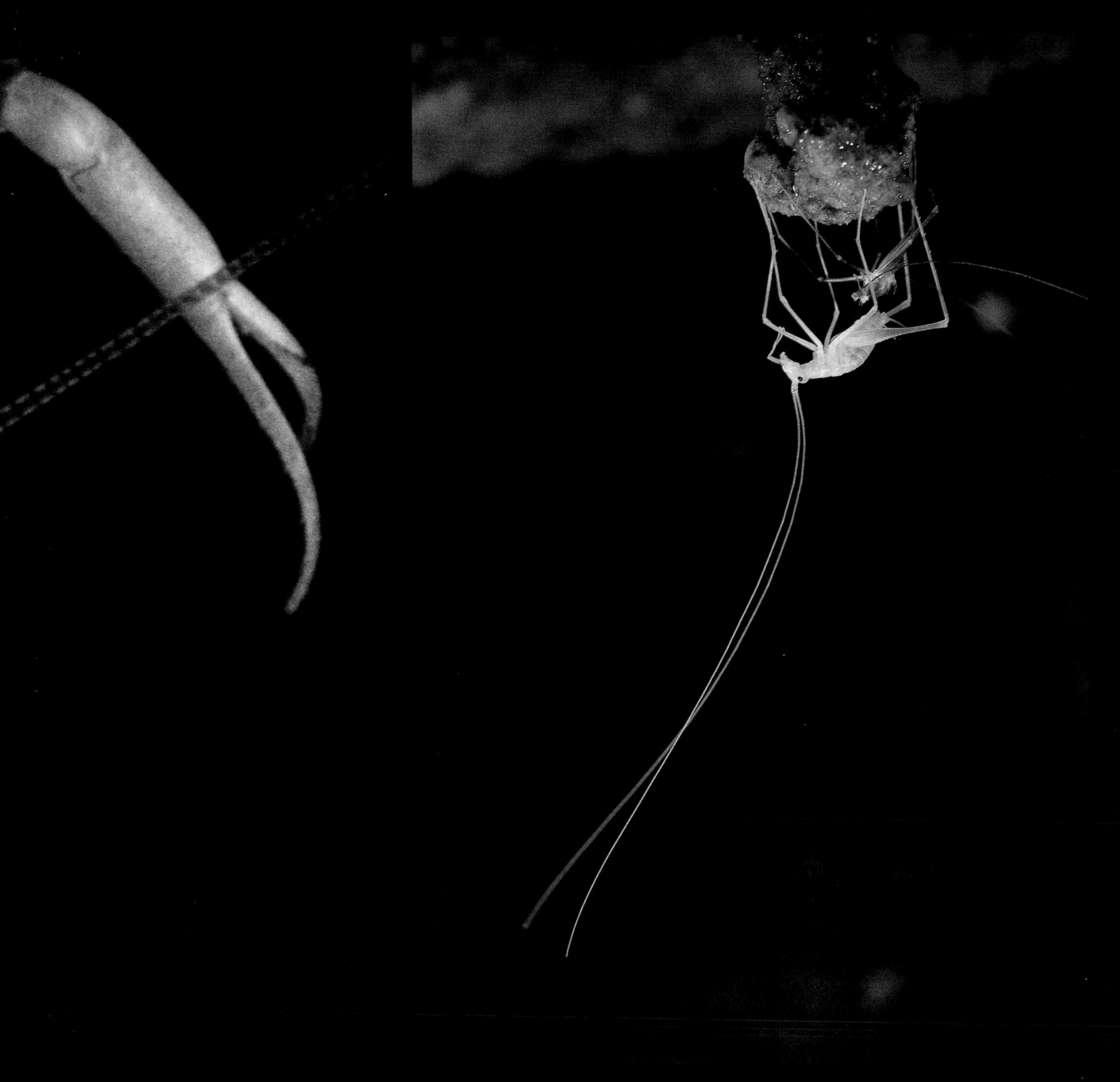

A cave cricket, clinging to a cavern ceiling with stiltlike legs, displays antennae that are twice as long as those of surface crickets. Unlike its chirruping relatives, the cave cricket is mute.

A close-up view of an inch-and-a-half-long, translucent cave crayfish shows the unique hairy feeding appendages that jut forward from its head. These oversized limbs are used to brush food, collected on the long legs and antennae, into the crayfish's mouth.

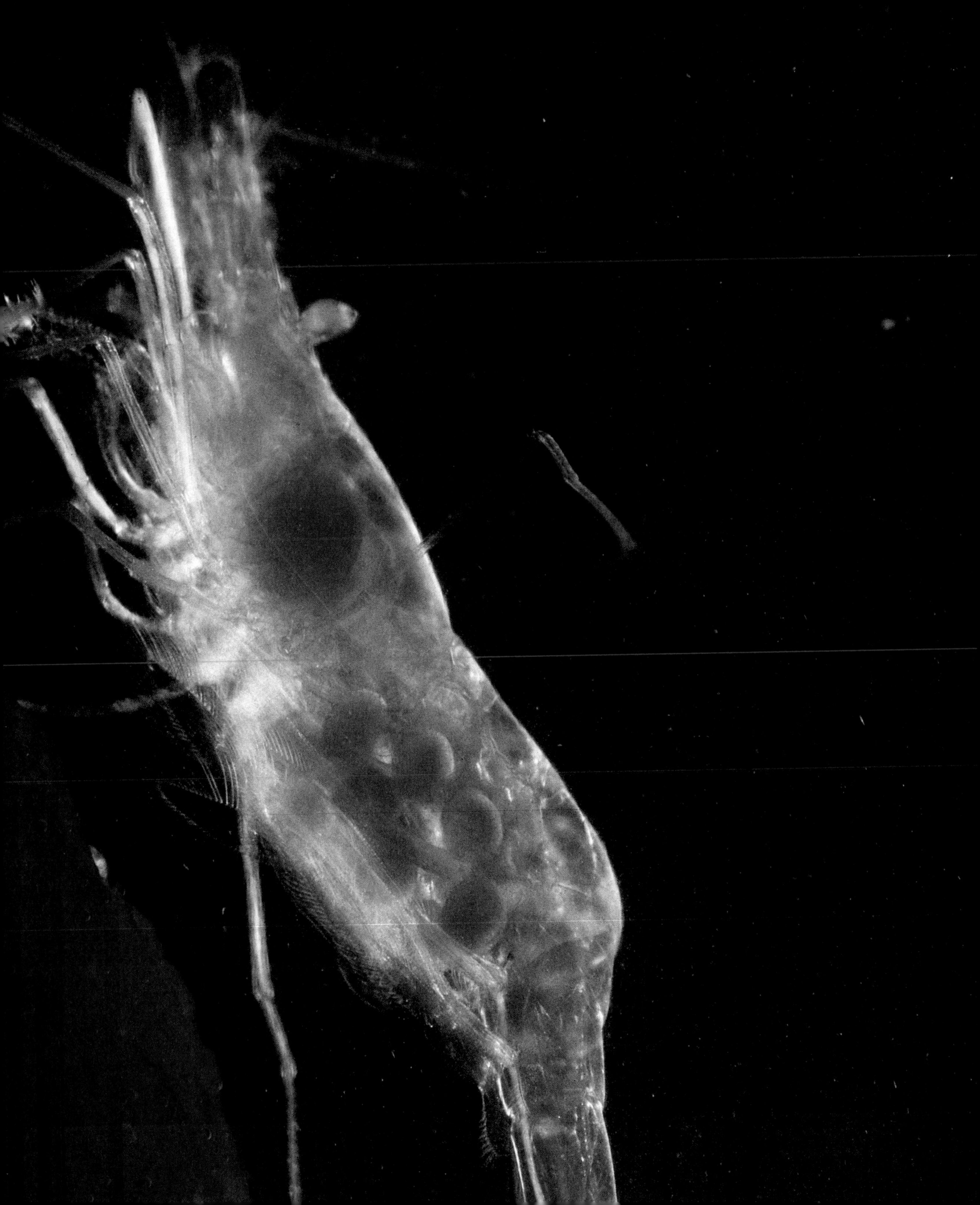

This rare blind shrimp is found only in Squirrel Chimney, a flooded sinkhole near Gainesville, Florida. The cave-dwelling shrimp produces fewer eggs than other shrimp varieties, but because the yolks are larger, the embryonic blind shrimp are well nourished and well developed when they hatch.

Even inside her web, a troglobitic spider keeps her eggs on her back. Most spiders living aboveground deposit their eggs on a convenient surface, but closer guarding of the eggs is necessary in the subterranean world because cave predators are so thorough in their quest for food.

Chapter 5

THE SPORTING SCIENCE

The deep and lasting satisfactions of caving, shading as they do into incommunicable and even mystical realms, are ultimately very personal. And in an earlier time they were supremely private, because so few people were willing to venture into caves at all. A subterranean frontiersman such as Floyd Collins or Norbert Casteret typically explored alone, or with one or two trusted companions, making his way with the greatest caution, thrusting a candle or a kerosene lantern tentatively ahead into the blackness.

The achievements of these early explorers were limited, not by any faintness of heart, but by their methods and equipment. To make a vertical descent, most cavers anchored a hemp rope on the surface and climbed down hand over hand, trusting their lives to their own strong arms and shoulders. Long ladder climbs were hardly less taxing; after venturing far into a cave, the fatigued explorer still had to haul himself back up to the comforting sunshine.

The benefits of 20th Century technology greatly increased the effectiveness of the individual caver. Instead of lighting his way with a flickering candle or lantern, he adopted the helmet-mounted carbide and electric head lamps favored by miners. Hemp ropes, easily frayed by sharp cave rocks and subject to rotting in the damp underground environment, were replaced by strong nylon lines, and ladders made entirely of rope gave way to flexible aluminum rungs suspended on steel cables. In Europe, mechanical winches were used for long vertical descents, but bulky machinery proved difficult to transport into the wilderness areas where most caves were to be found.

Perhaps the most important caving breakthrough came when cavers began to make use of the techniques and hardware perfected by mountaineers. The first such adaptation—for vertical rope ascents—was a knot designed for avalanche rescue by Austrian mountaineer Karl Prusik around 1930. The prusik knot slid easily upward along a rope, but when downward pressure was applied the knot tightened enough to bear a climber's weight. By attaching foot loops to a vertical rope with a prusik, the climber could walk slowly up the rope, sliding the knotted foot loops upward one at a time. Later, in caving as in mountain climbing, mechanical ascenders replaced the prusik knots, and a device called a rappel rack was invented to allow controlled one-rope descents.

Early in the 20th Century, serious cave explorers began to realize that they needed to improve not only their equipment but their teamwork if they were to advance the subterranean frontiers. In order to share knowledge and techniques, European and American cavers organized themselves

Members of an ill-fated 1952 expedition to the Cave of Pierre St.-Martin in the French Pyrenees watch caver Marcel Loubens begin the descent. The winch and cable under the shelter helped the team reach the floor of the chasm 1,135 feet below, but proved to be Loubens' nemesis.

Marcel Loubens *(center)* enjoys an animated chat with a Pyrenean shepherd *(left)* during an aboveground respite in the 1951 explorations of the Cave of Pierre St.-Martin. Listening in are Georges Lépineux, the cave's discoverer *(right)*, and expedition member Haroun Tazieff.

into clubs and federations. Caving societies began to blossom throughout Europe, and in 1941 the National Speleological Society was chartered in the United States. Within a decade, well-organized expeditions had replaced the solitary adventurings of the previous generations.

These expeditions were both systematic and purposeful. Using compasses to record the direction of movement, clinometers to measure angles of descent and ascent, and altimeters to determine depth, the teams of explorers and scientists meticulously charted every passage they found. Expeditions returned to the same cave year after year to probe promising leads.

The improved organization, methodology and technology of caving amounted, by 1950, to a thorough transformation of the art of underground exploration. It was a transformation that would make dramatic contributions to scientists' understanding of how caves form, how they change and how they have come to be the repositories of some of nature's most dazzling geologic treasures. At the same time, the known boundaries of the world's underground environments would be greatly expanded, most notably in such sprawling cave networks as Hölloch Cave in Switzerland and Kentucky's Flint Ridge system. Inevitably, these successes further whetted cavers' appetite for the singular combination of danger, beauty and conquest offered by the underworld frontiers.

The expeditionary approach to caving was still in its infancy in 1951 when 12 European cavers gathered in a remote section of the French Pyrenees for a team venture into an intriguing sinkhole. The previous year, one of their number, Georges Lépineux, had been searching for caves in the limestone mountain range when he spotted a crow emerging in full flight from what appeared to be a blank rock face. Looking closer, Lépineux found a small, concealed opening to a seemingly bottomless pit. He contacted his friend Max Cosyns—a Belgian scientist and a veteran caver—and Cosyns organized the assault. Among the men Cosyns enlisted were French volcanologist Haroun Tazieff and Marcel Loubens, a 28-year-old protégé of the eminent speleologist Norbert Casteret. For the initial descent into the shaft, Cosyns planned to use a pedal-powered winch and more than 1,300 feet of 1/5-inch steel cable.

Lépineux's sinkhole, called the Cave of Pierre St.-Martin after an old border post nearby, opened onto a rugged, wind-swept ridge at an altitude of 5,660 feet near the Spanish border. Here, long ago, the limestone rock had been compressed and folded by titanic mountain-building forces into a landscape riddled with deep vertical faults. Over the years, the substantial rainfall of the region had carved out spectacular chimneys leading to large underground chambers. Through these channels still coursed millions of gallons of water daily, bursting in powerful cascades from the cliffsides of nearby gorges.

Cosyns' group set up camp on the barren hillside and anchored the winch near the entrance to the shaft. Lépineux, as the discoverer, was the first to descend. Wearing a tight, rib-bruising parachute harness attached to the cable by a single metal clamp, Lépineux backed slowly over the edge. The crewman on the winch pedaled steadily, paying out the wire. Soon, Lépineux passed beyond earshot, and thereafter relied on a telephone line for communications with the surface.

Walking slowly backward down the wall of the shaft, Lépineux found a sloping outcrop ledge at a depth of 260 feet. Some 440 feet farther on, he descended through an icy waterfall that spilled out of a fissure in the wall. Drenched and bone-cold, he continued the descent, loosing showers of rocks as he negotiated a series of outcrops. After about an hour and a half, he found himself twisting slowly in open space as he was lowered through the roof of a large chamber whose floor appeared to be about 150 feet beneath him.

His chest aching from the brutal pressure of the harness, Lépineux finally touched down more than 1,100 feet below the surface—a new world's record for vertical descents. He scrambled down a rocky slope and explored the chamber for several hours, until fatigue drove him back to the winch line for the tedious return journey to the surface.

The next day, Tazieff and Loubens went down and started to explore in earnest. Beyond the area Lépineux had seen, they came to a small hole in the floor of the chamber; through it, they could see a terrace and another pit. Loubens explored the lead briefly and announced that a great chasm had been found, but this route down was too dangerous. Hours later, they found a safer descent. Loubens tied himself to a rope and climbed down into the passage, using the wire ladder they had brought along. Soon his voice rose from the gloom: "It's huge, positively huge. I can't see the walls. I'm going down a little farther."

Tired, alone in a darkness that seemed to swallow his feeble light, Tazieff settled back to wait, knowing that they must report back to the surface by phone in less than three hours. After a very uncomfortable hour, Tazieff shouted into the hole. There was no response. Another anxious, silent hour crept by. Finally, he heard and joyfully responded to a faint cry from Loubens. Both of Loubens' lights had gone out, but with frequent shouts and a brilliant magnesium flare, Tazieff managed to guide the exhausted but exhilarated explorer back. "Lost myself," Loubens explained nonchalantly, adding, "This is a *cave.*" A little while later, the tension and fatigue caught up with him. While rolling up the ladder after a climb, Loubens suddenly broke down and wept. "I was really frightened," he confessed.

Although he had flirted with disaster by going on alone, Loubens had found not only a large gallery, but also a caver's most sought-after prize—a river that undoubtedly led to more cave. Bearing this exciting news, the two men hurried back to the shaft, and soon afterward they were on their way up and out. Encouraging though their tidings were to the other team members, Cosyns declared that the winch was about to give out and called a halt to the expedition for that year.

The following summer, most of the original party returned, including Loubens and Tazieff, and this time they were joined by the French speleologist Norbert Casteret. Plans called for the men to be lowered into the pit quickly with an electric winch, but the machine was beset by gremlins, and delays were frequent. Loubens and Tazieff spent the whole of the first day getting to the bottom of the shaft and setting up camp. The following day, they surveyed the chamber that Loubens had discovered and saw that the river disappeared through an impassable opening in the rock. After a 10-hour search, they finally found a shaft that revealed a second large chamber below. Too tired to go on, they returned to camp, where they were joined by two other team members. With these reinforcements, they found a passage around the river siphon on the third day, but again exhaustion caught up with them before they could explore farther. That evening, Loubens announced that he would return to the surface to give someone else a chance. He phoned Casteret. "I've had my share of fun," Loubens told him. "I'm all in."

In the morning, the men helped Loubens into his harness and watched as he trundled up the pile of boulders they used as a lift-off platform. A few minutes later the cable stretched taut and he began to rise, spinning slowly. When he was about 35 feet up, he tried to ignite a flare so that Tazieff could take a photograph, but the matches blinked out one by one in the strong draft.

Still hoping for a picture, Tazieff was squinting up through the camera's viewfinder when he saw the beam of Loubens' head lamp suddenly plummet toward him. At the same time he heard a cry. An instant later, Loubens struck the boulder Tazieff was standing on and rolled down the rock pile, bouncing from stone to stone. Another team member finally halted his fall 100 feet farther down the slope.

Loubens was unconscious, breathing with rapid gasps. With infinite care on the perilous rubble slope, the men eased him onto a sheet of canvas and carried him back to the campsite, the only place where he could lie flat. While Tazieff removed Loubens' helmet and examined his skull for signs of fracture, one of the men scrutinized the dangling cable. At the end of the line he discovered the cause of the fall: The metal clamp that attached the harness to the cable had worn through.

They immediately phoned the expedition doctor, André Mairey, on the surface. He said that he would come down as soon as the cable was hauled up and the shackle repaired. In the meantime, the others could only wait and watch their friend's torturous efforts to breathe. Occasionally, they wiped his face with a damp cloth. Hours passed with no sign of Mairey, and then the phone inexplicably went dead. The surface crew labored to repair both the clamp and the phone, but progress was agonizingly slow. That night, a thunderstorm struck the topside camp with powerful winds, further delaying the doctor's descent. When Mairey finally reached the bottom with a stretcher, Loubens had been fighting for breath for nearly 24 hours. The doctor leaned over and examined him carefully. Loubens had a broken back and a fractured skull: His chances of survival were infinitesimal. But as long as he remained alive, the men were determined to try to get him out.

While Mairey and the men below strapped Loubens into a harness and bound him tightly to the stretcher, the topside crew linked together several 65-foot-long ladders and dropped them into the pit. Casteret and four others climbed down the ladders and took up positions at different depths on narrow ledges, anchoring themselves with pitons. The lowest man stationed himself 790 feet below the surface. Each was prepared to risk his life to shepherd the stretcher upward past the outcrops in the shaft. Far below,

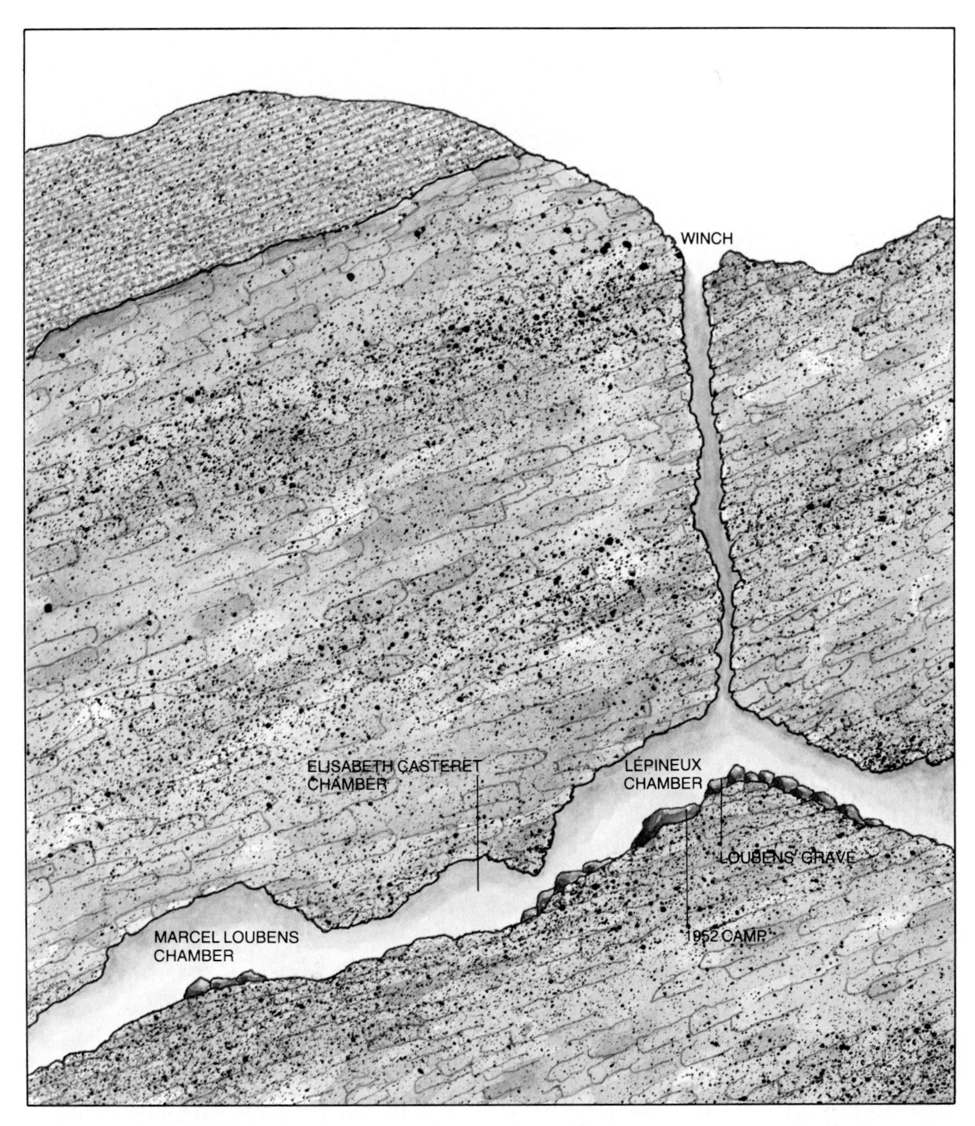

In the Cave of Pierre St.-Martin, a 1,100-foot shaft opens into a series of spacious underground chambers. After Marcel Loubens' fatal fall at the bottom of the shaft, his companions entombed him near their subterranean camp in a crypt of loose boulders.

rocks dislodged by the rescue team peppered the camp as Mairey gave the injured man a blood transfusion.

Mairey was still awaiting the go-ahead signal when he heard Loubens groan. Seconds later, almost 36 hours after his fall, Loubens' labored breathing stopped. The surface team received the solemn announcement by phone: "Marcel Loubens is dead." The word was passed to the men in the shaft, and one by one they climbed back up. The risks they were prepared to take to save an injured comrade were deemed too extreme simply to remove his body.

The next day, the men in the cave wrapped their friend's body in canvas and laid it in a hollow between two boulders, covering it with gravel and stones. Tazieff fashioned a small cross out of sheet metal. Then, using a hammer and chisel, the men carved an inscription into the rock near the grave: "Here Marcel Loubens passed the last days of his gallant life."

Although heartsick, Tazieff and Mairey decided to stay below long enough for a last look at the passage Loubens had found at the end of the second chamber. A series of twisting corridors brought them again to the elusive river. Making their way through a small room, they emerged into the main hall, a huge, smooth-floored vault decorated with soda-straw stalactites and large mushroom-shaped stalagmites. Tazieff lighted a flare and marveled at the chamber's grandeur—it was the largest room yet. At its low end, the room pinched down to a flooded tunnel. Here they decided to turn back, but first they named the spectacular cavern Loubens Chamber.

Cavers and bystanders cluster on the mountainside above the entrance to the Pierre St.-Martin system in 1952 to celebrate a mass in memory of Marcel Loubens. His death a week earlier at the bottom of the shaft curtailed the preliminary exploration of the cavern.

Casteret, at the age of 56, spearheaded the underground explorations when they resumed in 1953. Teams led by the veteran speleologist found four more large chambers. The last of the four measured some 660 feet by 390 feet, with a ceiling 330 feet high—the largest cave chamber in Europe. It ended at a flat and fissureless wall where the river, now merely a trickle, seemed to vanish into the floor. According to the cavers' altimeter, the gallery was deeper than any yet discovered—2,389 feet below the surface.

But for the Pierre St.-Martin explorers there remained the melancholy task of bringing Loubens' body back to the surface. The following year, they placed his remains in a coffin-like aluminum container and, wrestling it past the outcrops in the shaft, managed to haul it up with a motorized winch.

The sustained teamwork at Pierre St.-Martin, though shadowed by the death of Loubens, remains a triumph in the annals of caving. The expeditions demonstrated the merits of organization, pioneered a promising vertical technique and explored deeper into the earth than anyone had before. The same team approach to caving was used by other groups that came to Pierre St.-Martin in the 1960s and 1970s and discovered two more routes beyond Casteret's last probe.

While Cosyns and Casteret struggled with their winches at Pierre St.-Martin, American cavers in the Appalachian Mountains were experimenting with different methods for vertical caving. Instead of moving the line itself up and down from above, they used techniques that allowed the individual climber to control his own movements on a stationary vertical rope. At first, the cavers tried the movable prusik knot for one-man ascents, and for descending used the rappel, a method of walking backward down a cliff face, controlling the rate of descent by paying out rope gradually through load-reducing knots and fittings.

More imagination was required for descending into pits when contact with the walls could not be maintained. The solution was the rappel rack, a device that permits the caver to control a descent by weaving the rope around several horizontal braking bars to increase friction. The metal bars are movable: To descend faster, fewer bars are used or the spacing between them is increased; to go slower, the bars are pushed together or more bars are used to increase the purchase on the rope. American tinkerers also developed mechanical versions of the prusik knot. Like the knot, these mechanical ascenders clamp the rope tightly when bearing weight, but they slide upward when the weight is removed.

Vertical techniques were improved and refined in North America for almost a decade before an opportunity arose to give them a truly dramatic field test. In the mid-1960s, a group of American cavers took their gear to Mexico's Sierra Madre Oriental, site of a 1,092-foot hole called the Sotano de las Golondrinas. Here, they would have to move down and up a rope near the middle of the shaft, far from the walls. Using rappel racks, they made the spectacular "free-fall" descent in only 30 minutes, less than one third the time taken by the men on the winch in Pierre St.-Martin. Although the climb back up with the mechanical ascenders took about two hours, the cavers were able to rest en route; they used a third ascender attached to a chest harness to keep themselves balanced and upright while they caught their breath.

The technical innovations had made possible an entirely new kind of caving experience. As one Golondrinas veteran remembered: "I was suspended in a giant dome with thousands of birds circling in small groups near the vague backcloth of the far walls. Moving slowly down the rope, I had the feeling that I was descending into an illusion and would soon become part of it as the distances became unrelatable and entirely unreal."

Not all the changes taking place in speleology were dramatic: One of the most important trends was a new appreciation of carefully designed research plans and systematic surveys. The first great proponent of this approach was Swiss geologist Alfred Bögli, who supervised the investigation of Switzerland's Hölloch Cave in the early 1950s.

Hölloch Cave wanders for miles beneath the Swiss Alps, and for many years its only known entrance was near the bottom of a 2,800-foot series of passages and jagged chambers. Although Hölloch was discovered in 1875, only four miles of it had been traced when Bögli began surveying the area in 1945. Six years later, he became scientific director of a group exploring the cave. By then, a large iron gate had been built across the entrance to keep out interlopers, a move Bögli had reason to regret the very next year. In August of 1952, a flood marooned him and three teen-age assistants in the subterranean depths. "We were in a room with no flat place to lie down on, and the temperature was 41° F.," he recalled. For nine cold and tedious days, they rationed their food—"600 calories a day, mainly soup and bread"—and waited for the water to recede. On the 10th day, Bögli felt a draft, a sign that water no longer completely filled the passageway. He and the three boys navigated a siphon while holding on to a rope and made their way back to the entrance, where they found that their would-be rescuers had gone home and left the gate locked. They had to force the gate open to get out.

After lying for two years in the chill depths of Pierre St.-Martin, Marcel Loubens' body is hauled to the surface in 1954. Because of frequent snags on the walls, winching the aluminum coffin up the shaft took 20 hours.

Undaunted by the incident, Bögli returned each year to Hölloch during the one fortnight in December when the cave was dry enough for exploration. By 1955, he and his crews had mapped some 34 miles of the cave.

Bögli was thus understandably surprised when he read a newspaper article that year proclaiming a 23-mile Kentucky cave under Flint Ridge to be the longest in the world. He informed the press of the error, and members of the National Speleological Society in America soon contacted him to learn the details of his work. When they read his report, they conceded without protest. From then on, a lively and mutually supportive correspondence linked the two groups.

Bögli continued his methodical surveys with the aid of a small army of youthful assistants. Camping underground and working in relays, teams of researchers studied the geology and hydrology of Hölloch Cave as they continued their mapping. Their studies revealed that the oldest part of Hölloch had been formed more than a million years ago. The water that shaped the cave left scalloped wall formations clearly indicating the rate and direction of its flow. By the end of the 1970s, they had extended the underground frontiers of Hölloch Cave to more than 78 miles. And Bögli had poked into almost every recess himself; only a few proved off-limits to him. "We now have a unit of measurement called a 'Bögli,' " he said in 1981. "It signifies a space large enough for me to get through."

Although Bögli's early contact with the National Speleological Society had a somewhat deflating effect, it also sharpened the Americans' competitive instincts. And they had a glorious goal in mind—finding a connection between the caves under Flint Ridge and those of the nearby Mammoth Cave system. If successful, they would confirm the existence of a cave at least twice as long as Hölloch.

In the early 1950s, the Flint Ridge cluster of three large caves was known to extend beneath an area of roughly seven square miles. These caves—Salts, Crystal (formerly Great Crystal) and Colossal—had never been found to be connected, but together they accounted for many miles of known passages.

A mile to the south, on the other side of Houchins Valley, lay the great network of Mammoth Cave, carved out of a similar sandstone-capped limestone formation. The exact length of Mammoth Cave was not known; promotion brochures claimed 150 miles, but cavers knew that to be a wild exaggeration. Nevertheless, like the Flint Ridge system, Mammoth Cave contained scores of promising leads. Moreover, the floor of the valley separating Mammoth Cave Ridge from Flint Ridge was limestone, indicating the possibility of a link.

In an earlier time, such speleological sages as Édouard-Alfred Martel and Floyd Collins had surmised the existence of a connection, but neither had been able to find it. The Kentucky Cave Wars beginning after World War I made freelance exploration in the area dangerous, and new discoveries were invariably kept secret. In 1941, another complication arose: Most of the land in the area was declared a national park, and exploration was forbidden without official permission from the National Park Service.

But there was no keeping cavers from what they regarded, in a mixed but revealing metaphor, as the Everest of Speleology. In 1954, the National Speleological Society launched a major expedition to explore the caves under Flint Ridge. Beginning in the privately owned Crystal Cave, the 64-member expedition mapped several new passageways, some of which extended under the forbidden Park Service territory, but the group proved too large to accomplish much more. The operation ended after only one week.

Ironically, the next significant find was made not by an elaborate expedition, but by a pair of veteran cavers operating in the old-fashioned Kentucky style—on their own, carrying everything they needed. William Aus-

The Redoubled Risks of Cave Diving

However tantalizing the thoughts of what lay beyond, early cave expeditions were often thwarted by a submerged passageway, or sump. Only a very few daring cavers would dive into the black water in the hope that they could find their way to breathable air—and back.

The scuba equipment and insulating wetsuits developed for military use during World War II enabled cave divers to explore even long submerged passages with relative ease. In Cocklebiddy Cave in Western Australia, a 1979 expedition investigated a two-mile underwater corridor in dives lasting up to seven hours. And in southern France, cave divers reached a depth of 276 feet in their search for the still-hidden sources of the Fountain of Vaucluse.

For all its rewards, cave diving remains the pursuit of a passionate few willing to face the combined hazards of ordinary diving and of caving. Murky cave waters often defeat the most powerful lights and cause divers to lose their way in labyrinthine corridors; and intrusive cave formations pose a constant threat to a diver's air supply. One statistic defines the dreadful risk: Between 1960 and 1980, the submerged caves of Florida claimed the lives of 234 divers.

Crowned with lights and bristling with equipment, two divers prepare to investigate a submarine cave in the Bahamas. When exploring underground chambers, divers sling their air tanks at their sides for better maneuverability in tight spots.

His light probing the gloom, a caver glides through a sump in Boreham Cave, Yorkshire. The line suspended from the ceiling was positioned there during an earlier dive to be used as a guide for later explorations.

tin—whose family owned Crystal Cave—and his partner Jack Lehrberger were forever rambling on Flint Ridge in search of new caves. Much of the ground the two men covered lay inside the vast tract of Mammoth Cave National Park—where exploration was still forbidden. It was with understandable discretion, then, that Austin and Lehrberger in 1955 investigated a cave entrance marked "unknown" on the park map. Preliminary surveys led them to guess that Unknown Cave—as it came to be called officially—connected with nearby Salts Cave. Instead, the several miles of passageways they explored brought them to a remote section of Crystal Cave. Despite a poor guess, they had made an important connection.

Double-checking every cave survey map they could get their hands on, Austin and Lehrberger discovered that a branch of Unknown Cave passed under the Austin family's land. Bringing in a rock drill, dynamite and student volunteers from a nearby college, they soon blasted a passage into this branch. Thereafter, they continued their explorations from their home ground, circumventing Park Service regulations. By the end of the year, they had explored 23 miles of passageways. Their announcement claiming that the Flint Ridge system was the longest in the world was the one that brought them into contact with Alfred Bögli.

Participants in the lavishly equipped, 64-member Crystal Cave expedition of 1954 gather near the cavern's entrance along the route of their communication lines. Later explorations of the Flint Ridge caves proved that smaller, less cumbersome parties were more productive.

Emboldened by their success, Austin and Lehrberger joined several other experienced cavers to form the Cave Research Foundation, professing scientific aims in an attempt to legitimize their efforts. The Park Service soon decided that an organized group of disciplined, experienced cavers could be a great asset to the park, and in 1959 the foundation members were granted full access for exploration. With this franchise, they revived the expedition approach, and it paid off almost immediately.

Lehrberger and two other explorers were dispatched in 1960 into Colossal Cave. They quickly pushed beyond the known trails and followed a stream to a passage so low that they had to crawl on their bellies through the cold water. However, a little farther on, they climbed a ledge and found themselves in Salts Cave—thereby establishing another new Flint Ridge connection.

According to a survey map, only 160 feet separated the Colossal-Salts paired system and the Unknown-Crystal pairing. In the summer of 1961, two other Cave Research Foundation explorers felt a draft coming through a rock pile in Unknown Cave and clawed at the rubble until they made an opening large enough to crawl into. They struggled through and emerged 30 minutes later in Salts Cave, thus demonstrating that all four Flint Ridge caves were in fact one 30-mile-long system. The explorers once again cast longing looks across the broad Houchins Valley toward Mammoth.

During the next three years, teams of cavers systematically explored the avenues of the Flint Ridge caves that led toward the valley. Finally, in 1964, they discovered a long crawlway that projected under the north end of the valley. Several exploratory missions extended this trail to an area only 2,000 feet from Mammoth Cave.

The following year, another team pushed on toward Mammoth until they were stopped by a massive breakdown pile beneath Mammoth Cave Ridge just 800 feet short of the cave. Team members tried repeatedly to clear an opening through the breakdown, but every time they removed one rock, another tumbled into its place. Finally, they gave up, and the hope of using this route to unite the two gigantic cave systems faded and all but flickered out. Only a special kind of speleological stubbornness kept it alive.

The Flint Ridge scramblers continued to probe every opening they could find. None paid off. It was scant consolation that by 1969 their dogged surveys had earned the Flint Ridge system the title of the world's longest cave, surpassing both Mammoth and Hölloch.

In May 1972, six years after the last attempt at the boulder choke, they returned to the clogged passage. Under the leadership of John Wilcox, a 35-year-old mechanical engineer from Columbus, Ohio, the Cave Research Foundation teams again attacked the exasperating wall of boulders. They hacked away for hours with digging tools, but got nowhere.

In July, however, team member Patricia Crowther found something new. A reed-thin physicist and the mother of two children, Crowther was already an experienced rock climber and caver when she joined the Flint Ridge group. Several hundred feet back from the rock pile, she crawled into a horizontal slot in an area that had not yet been thoroughly probed. After squirming about 100 feet, she came to a narrow, wedge-shaped crawlway, soon to be known as the Tight Spot. Pushing her back against its low ceiling so she would not slip into the V-shaped fissure beneath her, she hauled herself through and emerged at the top of a small waterfall that spilled over into a stand-up chamber. Her compass showed that the water was draining west toward Mammoth, and a lead was in sight. A few minutes later, she crawled back and rejoined the others. "It's very tight," she announced. "But we have cave!"

More than a month passed before Wilcox was able to send a small team back to look at Crowther's discovery. Veteran speleologist Roger Brucker, his 19-year-old son, Tom, and Flint Ridge newcomer Richard Zopf made their way to the Tight Spot. The elder Brucker was unable to get through, but Zopf and Tom Brucker made it. While Zopf caught his breath, Tom hurried through the stand-up room and slogged some 1,000 feet along the stream in a mud-walled passage. He felt certain that the stream emptied into the Mammoth Cave system, but he decided to wait for a full team before proceeding farther.

Four days later, Wilcox, Crowther, Zopf and Tom Brucker returned to

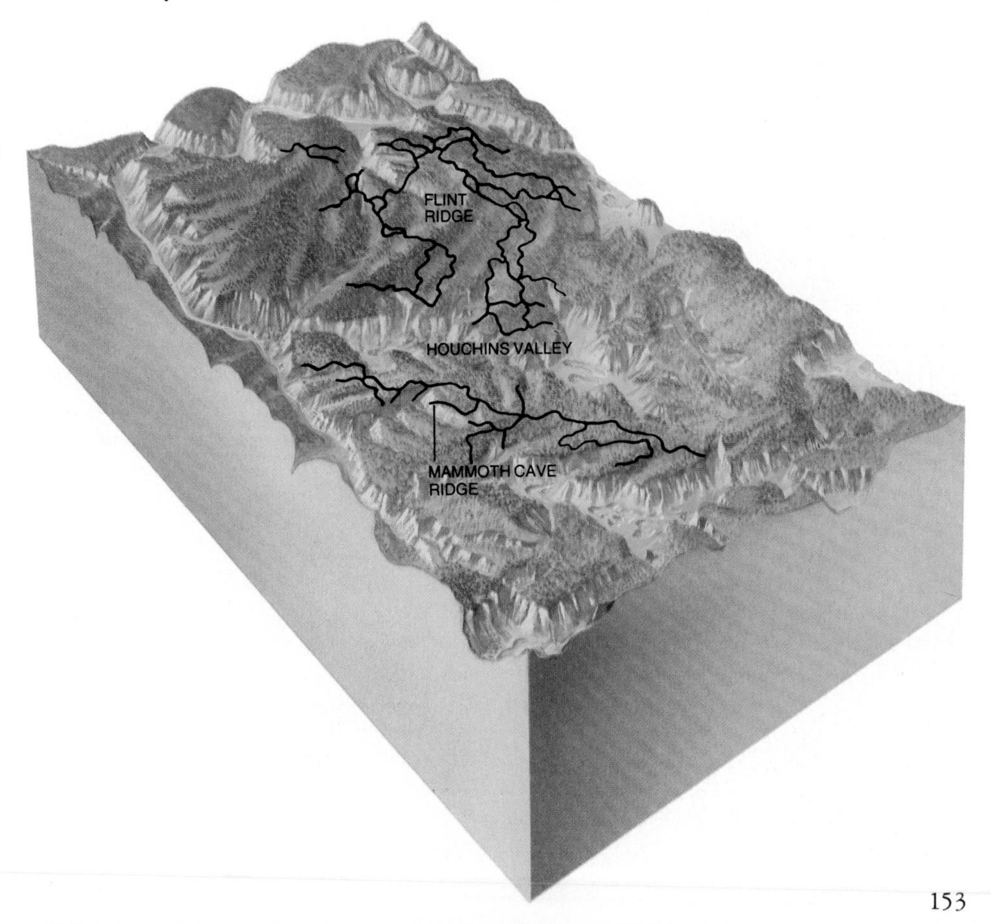

By 1961, as shown in this relief view, cavers had explored elaborate networks of passages beneath the uplands of Flint and Mammoth Cave Ridges. But they knew of no route into the limestone bedrock of Houchins Valley.

the passageway. Beyond the Tight Spot they began a systematic survey of the tortuous route Tom had found. By 9:30 p.m., nearly 12 hours after setting out, they reached the point where Tom had turned back. Knowing that the return trip would take at least six hours and that a rescue team had orders to start in after them at 4 a.m., they decided to explore for only 15 minutes more.

Tom led the way ahead. The passage narrowed down to a 50-yard section that he navigated on his belly. On the other side he found a stand-up passage—and there spotted an arrow carved in a mud bank. By the time the others joined him, he had found the name "Pete H." and the initials "P.H." and "L.H."

They all knew who P.H. and L.H. were: Pete Hanson and Leo Hunt had been guides in Mammoth Cave in the 1930s. Both were dead, but it was known that they had explored passages branching off the Mammoth rivers. Conceivably, of course, the two guides had reached this point from a sinkhole that was not connected to Mammoth, but the team was too excited by the prospect of an imminent victory to entertain doubt.

They drove on through water that was knee-deep, then waist-deep. The muddy walls were evidence that the passage was regularly flooded by the fall rains. When they spotted a small brown fish with eyes, they knew that they had to be close to the Green River, which flowed on the surface near Mammoth Cave Ridge.

Just before midnight, they crawled into a three-foot-high passage and rested on a patch of dry ground. They estimated that they had traveled five and a half miles in all, and although the stream flowed on ahead, it was time to turn back. Wilcox burned their initials on the wall with his carbide lamp. Sagging with exhaustion, they struggled out of the Austin entrance just after sunrise. The rescue team had decided to give them a little leeway before launching a search.

The summer caving season was nearly over. September rains might soon flood the cross-valley passage. If they were going to make the final connection that year, they would have to act quickly.

On September 2, Wilcox led another team down under the valley, but they were forced to stop short of their goal when one member could not get

Traced in the soft mud of a remote corridor by Mammoth Cave guides in the 1930s, initials and an arrow indicate the way toward the main chambers. A party of cavers reached the markings from the Flint Ridge cave system in 1972, then followed the arrow into Mammoth, demonstrating for the first time that the two caverns were connected.

through the Tight Spot. The next day, Crowther and three others donned wetsuits and plunged into Mammoth Cave's Roaring River, hoping to find Pete Hanson's old route from the other side. They shivered for hours in the 54° F. water, but found nothing.

Wilcox decided to try the Tight Spot again with a six-member team, including Zopf and Crowther. He assembled his supplies, alerted his backup rescue team and kept a keen eye on the weather. The rain held off. On September 9, the explorers ducked into the cave at 10:30 a.m. All six made it through the Tight Spot, and when they arrived at the end of the earlier survey, they unreeled their steel tape and began carefully measuring the newly discovered sections. After surveying some 4,000 feet, Wilcox gave the order to put away the measuring equipment and press on with the exploration.

With Wilcox in the lead, they proceeded slowly into deeper water. The ceiling lowered, dipping to within a foot of the green water. Wilcox moved ahead, hugging the wall. "Wait here," he told the others. "No point in everyone getting wet if it siphons." Waves stirred up by his movements lapped the ceiling. The others stood by and fretted. Soon Wilcox realized that the ceiling was receding again. He waded another few feet, and then the ceiling seemed to disappear. "I'm through," he shouted. "I've got something! It's big!" The others heard not only his words but something even more thrilling—an echo. The rest of the team sloshed forward.

As the others approached, Wilcox squinted ahead into the gloom. He saw a wall in the hazy distance and then something else—a metal handrail. "I see a tourist trail!" he cried. The others whooped and hurried to catch up; several stumbled in their haste. Crowther slipped and sat down in the water up to her neck. The new chamber, she recalled later, looked like "a moonless, starless night." All were euphoric. The culminating triumph gave Wilcox, "a feeling of vastness inside the skull." After almost two decades of patient, tough caving, the final connection had been made; Mammoth and Flint Ridge were one 144-mile-long cave. Meticulous to the last, Wilcox's team completed their survey of the link and then made their way out through one of Mammoth's entrances.

When Crowther returned a few months later, the connecting passageway

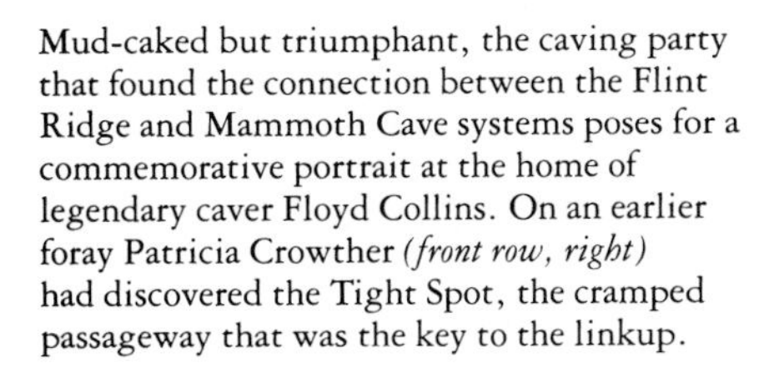
Mud-caked but triumphant, the caving party that found the connection between the Flint Ridge and Mammoth Cave systems poses for a commemorative portrait at the home of legendary caver Floyd Collins. On an earlier foray Patricia Crowther *(front row, right)* had discovered the Tight Spot, the cramped passageway that was the key to the linkup.

was flooded, as it is for much of the year. Almost certainly this explained how the route had been overlooked for so long. But later, when Stephen Bishop's map of 1842 was consulted, it showed this crucial passage had in fact been discovered by Mammoth's first and most famous guide.

As compelling as the adventure and romance of underground exploration may be, there are practical reasons for knowing the location and extent of caves. One of these was demonstrated by what happened to a cave just 10 miles east of the Flint Ridge system. Hidden River Cave, opened for tourism in 1916, boasted elegant galleries and offered scenic boat rides along the river's winding course through the cave. Hidden River also supplied drinking water to a small town situated above the cave; for decades the townspeople repaid this kindness by dumping sewage into nearby sinkholes, never suspecting that the sinkholes might feed back into the river. By the mid-1930s the water supply was hopelessly contaminated, and a decade later the tourist industry came to a halt as vile odors drove visitors back to the surface. The half-mile-long show cave and some 20 miles of lesser passages became a large sewer: The once-plentiful population of pearly blindfish in the river disappeared.

The Hidden River Cave disaster is by no means unique. Throughout the world, pollution from the surface poses a serious threat to caves—and thus to groundwater supplies at unexpected distances from the source of pollution. Karst regions are particularly susceptible to contamination because groundwater flows swiftly into the caves through joints in the rock instead of filtering slowly through purifying layers of sand and soil. The pollutants in the water typically sluice down into the cave's delicate ecosystem with all their raw toxicity intact. Cave creatures, which live precariously in the darkness, can be wiped out in a matter of hours.

Cave pollution has been a recognized problem ever since Édouard-Alfred Martel demonstrated in the 1890s that groundwater in karst areas carried disease-spreading bacteria and viruses. Martel's work to ban the dumping of wastes and animal carcasses into sinkholes undoubtedly saved thousands of lives in early-20th Century France, but underground water pollution still threatens residents of other European karst regions, especially in Austria, Yugoslavia and Italy, where large towns and cities tap subterranean streams for drinking water.

After World War II, the rapid rebuilding of European industry made cave pollutants even more deadly. Many industrial by-products are highly toxic and remain so for years; chemical wastes and heavy metals such as copper, mercury and chromium may poison the fragile underground environment for generations. In Yugoslavia, industrial pollution in karst regions annihilated more than 100 species of animal life in the famous Pivka underground river. More than a third of Austria is karstic, and in some places drinking water travels from a karst plateau to the household tap in less than half a day. Austria has designed special legislation to restrict development, industrialization and deforestation in karst regions.

A similar postwar industrial boom in the United States brought the same problems to American cave systems. And the pollution posed a frequent hazard for the steadily increasing number of amateur cavers. In 1966, the flame in a caver's carbide lamp ignited gasoline fumes in a cave about 100 miles northwest of Atlanta, Georgia, and the carbon monoxide fumes—formed by the incomplete combustion of the gasoline in the cave—killed three men. An investigation after the accident revealed that the gasoline had most likely seeped into the ground from a leak in a storage tank at a nearby service station. Three years later, a layer of gasoline seven feet thick

A map of the maze of the Flint Ridge and Mammoth Cave systems shows the route *(white)* of the expedition that proved they were connected. Although the charted passages appear to cross frequently, they actually run at many different depths and rarely intersect.

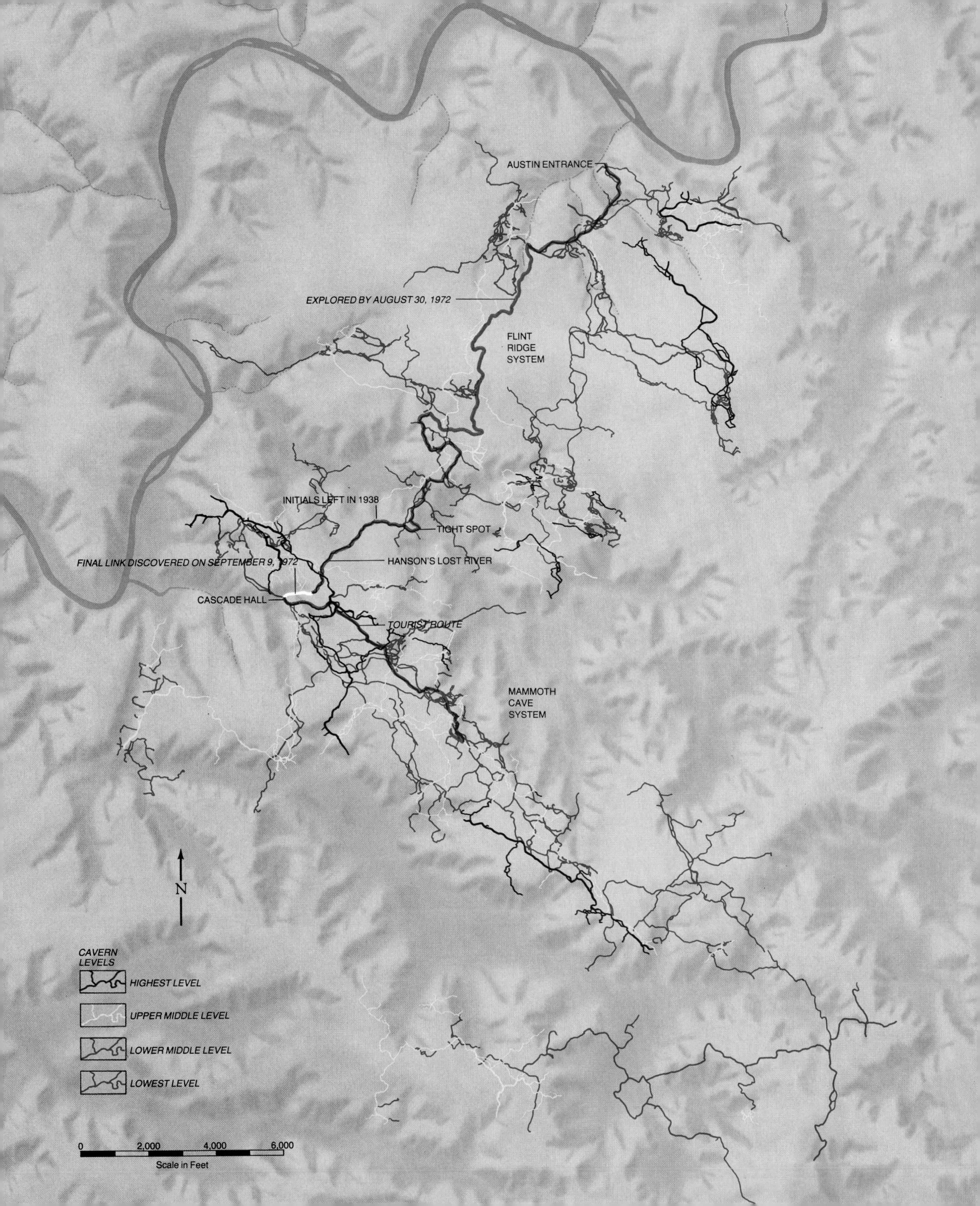
AUSTIN ENTRANCE
EXPLORED BY AUGUST 30, 1972
FLINT
RIDGE
SYSTEM
INITIALS LEFT IN 1938
TIGHT SPOT
FINAL LINK DISCOVERED ON SEPTEMBER 9, 1972
HANSON'S LOST RIVER
CASCADE HALL
TOURIST ROUTE
MAMMOTH
CAVE
SYSTEM
N
CAVERN
LEVELS
HIGHEST LEVEL
UPPER MIDDLE LEVEL
LOWER MIDDLE LEVEL
LOWEST LEVEL
0
2,000
4,000
6,000
Scale in Feet

was found floating on top of the groundwater in a well in a Pennsylvania karst area. The prime suspect was a local storage facility used by several oil companies. Almost one third of a square mile of groundwater was coated with gasoline; recovery crews collected more than 200,000 gallons.

As the world's karst areas become more populous, underground cavities can present another, even more direct threat to the surface dwellers. In a karst region in southern Italy, a house suddenly disappeared in 1978 when a cave roof collapsed beneath it, opening a new sinkhole. A parked car was sucked down with the house while the driver was away buying ice cream. In the United States, similar incidents occur fairly frequently in karst areas, where the land is honeycombed with subterranean cavities. In Alabama, 4,000 sinkholes have formed since 1900. Caused by drainage of underground water, these sinkholes have damaged highways, streets, railroads, buildings and pipelines. In central Florida, a sinkhole that developed in Winter Park in 1981 caused damage estimated at two million dollars.

Costly as these mishaps are, the effects could be far more disastrous if a sudden collapse destroyed a large dam. In any case, dams built on limestone are subject to other ills, including frequent leakage due to dissolution of the supporting rock. Repairs to dams in American karst regions have already cost tens of millions of dollars.

In an effort to detect subterranean cavities before major construction projects get started, scientists have developed a variety of geologic sensing devices. Among the most successful are underground radar, seismic detectors, acoustic resonance sensors and equipment that measures the change in the flow of an electric current around an underground cavity.

All these instruments rely on essentially the same principle—the measurement of changes in the flow of energy through the ground. Because the speed of energy traveling underground can be predicted accurately, changes in the rate can reveal the existence of a tunnel or a cave. For the seismic method, charges are detonated in boreholes to create measurable vibration. Radar waves transmitted through the ground between two sensors are equally predictable, and again, any interruption in the known pattern will indicate the presence of a cavity.

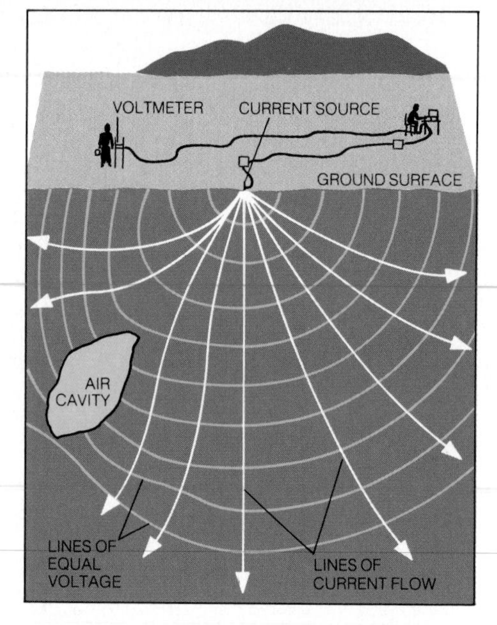

Technicians in a Texas field operate equipment that can locate underground voids by charting patterns of electrical resistivity. Current sent through the earth from a source electrode *(diagram)* loses voltage uniformly with distance unless an underground cavity disturbs the pattern; a computer can interpret the readings to plot the location, size and shape of the cavity.

To measure the extent of a cave that is already known, a high-intensity loudspeaker is sometimes placed inside the cave to blast sound into the cavity, causing the walls to resonate. This acoustic vibration travels through the rock walls to sensors on the surface. Analysis of the vibration then yields a statistical picture of all parts of the cave that connect directly to the sound source. For the detection of shallow cavities as small as several cubic feet, scientists measure the resistance of the soil to electrical energy. Electricity supplied by a portable generator is directed into the ground beneath a line of uniformly spaced electrodes. Because air is a poor electrical conductor, an empty cavity shows a significant resistance to the electrical current.

This new technology is showing promise for such applications as tracing karstic water supplies, and it might even be of use in finding new branches of major caves. But cavers by temperament tend to favor more traditional methods. Still working in teams and mapping carefully as they stoop, crawl and slither into new passages, explorers continue to push back the boundaries of the world's great caverns.

Kentucky cavers hardly paused for breath after the historic linkup of the Flint and Mammoth Cave Ridges. In 1979 the size of the Flint-Mammoth system was increased again when it was linked to the Joppa Ridge system southwest of Mammoth Cave. By 1982, the Flint-Mammoth-Joppa system had been extended to some 230 miles, and many tantalizing leads remained. "There is no doubt that it will reach 300 miles," said one Mammoth geologist, "and 500 is not impossible."

New records continue to be set in European karst regions as well. In 1981, a group exploring in the French Alps established a world depth record of 4,800 feet in a cave called Jean Bernard. After more than 30 years of exploration, Hölloch Cave in Switzerland is, at 87 miles, the third longest cave in the world, surpassed only by the Mammoth-Flint-Joppa system and the 89-mile Optimistichekaya Cave in the Ukraine.

Despite all the new detection technology, there is really no way to predict how many caves remain undiscovered. Most geologists agree that perhaps fewer than half of the caves in the United States have been located. With modern caving methods and equipment, it is difficult to imagine a cavern that could not be conquered by a crack team of speleologists. But in order to find these caves, explorers throughout the world will in the end rely on the most enduring caver's trait of all—passionate curiosity. Beyond that, they need a little luck and some well-earned experience. "I start with what I know and then poke around," said one veteran speleologist. "When I find a sinkhole I start digging." Ω

A "NEW CONTINENT" OF KARST

Despite the enthusiastic efforts of generations of cavers, only a small fraction of the earth's caves have so far been discovered, let alone explored. Yet while vast underground worlds remain hidden, speleologists have a good general idea of where they are. The caves lie beneath some of the earth's most distinctive and exotic terrain—areas of exposed limestone known as karst.

Named after the Karst area of Yugoslavia, where Serbian geographer Jovan Cvijič conducted the first comprehensive study of them in 1893, these landscapes are recognizable by the pronounced erosion of their highly soluble limestone. Rain water drains from other types of terrain into rivers and streams running through connected valleys. But rain falling on karst is absorbed almost immediately into the limestone's cracks, pits and sinkholes. Dissolution of the rock continues under the surface until networks of caves are carved out. The caves provide even better drainage and often cause rivers to disappear into the ground. Large depressions pockmark the karst landscape; the soil is thin or nonexistent; and, except in tropical regions, vegetation is sparse.

Karst regions occur throughout the world, notably in the Mediterranean basin, in parts of the Alps and the Pyrenees, in Kwangsi Province in southern China, and in Kentucky, Missouri and Tennessee in the United States. Local climatic conditions, especially rainfall, and details of geological history cause great variations in the appearance of karst, as the photographs on the following pages attest. The accompanying illustrations show where an experienced caver would go in pursuit of his ultimate achievement—the exploration of a previously unknown cave.

The frontier is so broad and the opportunities so immense that in 1966 the noted Swiss speleologist Alfred Bögli was moved to write: "Under the earth's crust there exists such an enormously great world, in absolute darkness, that we can with some justice speak of a new continent."

A karst formation called limestone pavement sprawls across the countryside in Yorkshire, England. The surface of the exposed limestone, scoured by glaciers during the last ice age, has been sculpted into deeply etched formations by water gradually eroding its joints.

This karst landscape, called a *polje,* from the Serbo-Croatian word for "cultivated field," is in the Taurus Mountains of southwest Turkey. Melting snow floods the flat-floored valley each spring; the rest of the year it is dry.

Most of the caves to be found in a *polje* occur along its edges, because alluvial deposits on the valley floor prevent absorption of the seasonal flood and the water drains away at the base of the limestone hills.

The limestone cliffs of a karst canyon loom nearly 3,000 feet over a dry riverbed in Vikos Gorge in northwest Greece. The river flows aboveground in the wintertime, but during the dry summer season it disappears into underground cavities, emerging at the lower end of the gorge as a great spring.

Cave systems are found in the walls of a karst canyon at successive levels of the water table. As the water table dropped, the older caves were left high and relatively dry while the cave-building process continued below.

Depressions surrounded by cones of limestone give cockpit karst, found only in the tropics, its unique appearance and account for its name. The jungle canopy of this Jamaican karst region softens its contours, but conceals a relief as jagged as that of other karst limestone.

In cockpit karst, the dissolution of limestone is intensified by the carbonic acid generated by the decay of the abundant vegetation. Most cave entrances are found beneath the cockpits; older caves higher in the cones were formed along previous water-table levels.

An imposing array of 650-foot-high limestone peaks overlooks fertile farmland in this example of tower karst in southern China's Kwangsi Province. Both the towers and the flat valley floor below them are products of thousands of years of erosion.

The ancient meandering river that is carving out a cave at the base of a limestone tower *(right)* formed the older, higher caves as it cut its way downward. At left, accumulating rock debris broadens a tower base to a more conical shape

ACKNOWLEDGMENTS

For their help in the preparation of this book the editors wish to thank: **In Austria:** Vienna—Dr. Hubert Trimmel, University of Vienna. **In Belgium:** Chaudfontaine—Raymond Tercafs, Films Astrolabe. **In France:** Aven Armand—Jean Cabantous; Carmetin—Patrick Pallu; Moulis—Christian Juberthie, Laboratoire Souterrain; Nice—Michel Siffre; Paris—Claude Chabert; Jacques Choppy; Marcel Ichac; Haroun Tazieff; St.-Gaudens—Norbert Casteret; Valence—Pierre Agéron; Le Vésinet—Jacques Ertaut. **In Great Britain:** Leeds—Dr. Barry Webb, Institute of Geological Sciences; North Humberside—Andy Eavis; Nottingham—Dr. Tony Waltham, Trent Polytechnic. **In Italy:** Castellana—Nicola Mongelli, Grotte di Castellana. **In the People's Republic of China:** Beijing—Dr. Song Lin Hua, Institute of Geography, Academia Sinica; Karl-Heinz Bernhardt, Yu Ming-Bao, Shanthi Wikkranasinha, Li Xiaocong, Peking University; Zhao Fang, New China Picture Company. **In Switzerland:** Zurich—Alfred Bögli, Silva Verlag. **In the United States:** Alabama—(Huntsville) Jeanne Pritmore, The National Speleological Society; (Tuscaloosa) John G. Newton, Water Resources Division, United States Geological Survey; California—(Los Angeles) Dr. Clement Meighan, Department of Anthropology, University of California; Connecticut—(Storrs) Karen Kastning, Department of Geology and Geophysics, University of Connecticut; District of Columbia—Sarah and William Bishop, Cave Research Foundation; Duncan Morrow, Chief of Media Information, National Park Service; Indiana—(Marengo) Gordon L. Smith, President, National Caves Association; (Shelbeyville) Harold Meloy, Mammoth Cave Historian; Kentucky—(Louisville) Sharon Bidwell, Bernice Franklin, Librarians, *The Courier-Journal* and *The Louisville Times;* Tom Owen, Assistant Director, University Archives and Records Center, University of Louisville; (Mammoth Cave) Lewis D. Cutliff, Assistant Chief Park Interpreter, Mammoth Cave National Park; Massachusetts—(Boxboro) Ann Kress, Chairman, National Speleological Society Photo Archives; (Wilbraham) Emily Davis Mobley, Speleobooks; Missouri—(Protem) Tom Aley, Ozark Underground Laboratory; New Jersey—(Closter) Russell Gurney; New Mexico—(Carlsbad) Bobby L. Crisman, Ronal C. Kerbo, Carlsbad Caverns National Park; New York—(The Bronx) Brother G. Nicholas Sullivan, Manhattan College; Ohio—(Dayton) Roger W. Brucker; Pennsylvania—(Altoona) Jack H. Speece, Editor, *Journal of Spelean History;* (State College) Dr. and Mrs. William White; Tennessee—(Knoxville) Bill Deane; Texas—(Austin) William Mixon, Editor, *Speleo Digest;* (Dallas) Pete Lindsley, President, Richard Zopf, The Cave Research Foundation; (San Antonio) Thomas E. Owen, Southwest Research Institute; Virginia—(Arlington) Chip Clark; (Luray) Robert L. Bradford, Kermit B. Cavedo, Luray Caverns; H. T. N. Graves, President, Luray Caverns Corporation; (Reston) United States Geological Survey Library; (Richmond) John Wilson, Virginia Cave Commission; (Springfield) Paul Stevens, Executive Vice-President, National Speleological Society; Washington—(Seattle) Dr. William R. Halliday; Robert Stitt; Wisconsin—(Milwaukee) Dr. Merlin D. Tuttle, Curator of Mammals, Milwaukee Public Museum. **In Yugoslavia:** Ljubljana—Grozdana Kosak, Graficki Kabinet, Narodni Muzej; Postojna—France Habe; Srečko Šajn, Dino Borsellino, Postojna Jama.

The editors also wish to thank the following persons: Janny Hovinga, Wibo van de Linde, Amsterdam; Bob Gilmore, Auckland; Enid Farmer, Boston; Brigid Grauman, Brussels; Robert Kroon, Geneva; John Dunn, Melbourne; Dag Christensen, Oslo; Jimi Florcruz, Peking; Eva Stichova, Prague; Mary Johnson, Stockholm; Traudl Lessing, Annelise Shulz, Vienna.

The index was prepared by Gisela S. Knight.

BIBLIOGRAPHY

Books

Ammen, S. Z., *The Caverns of Luray.* Allen, Lane & Scott's Printing House, 1886.
Bailey, Vernon, *Cave Life of Kentucky.* The University Press, no date.
Baker, Robin, ed., *The Mystery of Migration.* The Viking Press, 1981.
Balch, Edwin Swift, *Glacières or Freezing Caverns.* Johnson Reprint Corporation, 1970.
Balch, H. E., *Mendip: The Great Cave of Wookey Hole.* Bristol: John Wright & Sons, Ltd., 1947.
Baring-Gould, S., *The Deserts of Southern France.* Dodd, Mead & Co., 1894.
Barnett, John, *Carlsbad Caverns National Park, New Mexico.* Carlsbad Caverns Natural History Association, 1979.
Bauer, Ernst, *The Mysterious World of Caves.* London: Collins Publishers, 1971.
Bennett, Ross, ed., *The New America's Wonderlands: Our National Parks.* The National Geographic Society, 1980.
Bidegain, J., et al., *Marcel Loubens: Ses Souvenirs, Nos Témoignages.* Paris: Gallimard, 1958.
Binkerd, A. D., *The Mammoth Cave and Its Denizens: A Complete Descriptive Guide.* Robert Clarke & Co., 1869.
Bishop, Sherman C., *Handbook of Salamanders: The Salamanders of the United States, of Canada, and of Lower California.* Comstock Publishing Co., 1943.
Bögli, Alfred:
Féerie du Monde des Cavernes. Zurich: Editions Silva, 1976.
Karst Hydrology and Physical Speleology. Transl. by June C. Schmid. Springer-Verlag, 1980.
Boon, J. M., *Down to a Sunless Sea.* The Stalactite Press, 1977.
Borror, Donald J., *A Field Guide to the Insects of America North of Mexico.* Houghton Mifflin, 1970.
Brook, D. B., and A. C. Waltham, eds., *Caves of Mulu.* London: The Royal Geographical Society, 1978.
Brucker, Roger W., and Richard A. Watson, *The Longest Cave.* Alfred A. Knopf, 1980.
Bullit, Alexander Clark, *Rambles in the Mammoth Cave, during the Year 1844 by a Visitor.* Johnson Reprint Corporation, 1973.
Caiar, Ruth, and Jim White, Jr., *One Man's Dream.* Pageant Press, 1954.
Call, Ellsworth, *The Mammoth Cave.* Louisville & Nashville Railroad Co., 1856.
Casteret, Norbert:
The Darkness under the Earth. Henry Holt and Co., 1954.
The Descent of Pierre Saint-Martin. Transl. by John Warrington. The Philosophical Library Inc., 1956.
E.-A. Martel: Explorateur du Monde Souterrain. Paris: Gallimard, no date.
More Years under the Earth. Transl. by Rosemary Dinnage. London: Neville Spearman Ltd., 1961.
My Caves. Transl. by R. L. G. Irving. London: J. M. Dent & Sons, Ltd., 1947.
Ten Years under the Earth. Ed. and transl. by Barrows Mussey. Zephyrus Press, 1975.
The Caves of the Earth: Their Natural History, Features, and Incidents. American Sunday-School Union, 1953.
Charles, Jean-J., *Norbert Casteret.* Monaco: Éditions les Flots Bleus, 1958.
Chevalier, Pierre, *Subterranean Climbers: Twelve Years in the World's Deepest Chasm.* Transl. by E. M. Hatt. Zephyrus Press, 1976.
Clemens, Samuel, *Adventures of Huckleberry Finn.* Harper and Row, 1965.
Cullingford, C. H. D., *Exploring Caves.* Oxford University Press, 1951.
Darwin, Charles, *The Origin of Species by Means of Natural Selection.* Avenel Books, 1979.
De Joly, Robert, *Memoirs of a Speleologist; The Adventurous Life of a Famous French Cave Explorer.* Ed. by Pierre Boulanger. Transl. by Peter Kurz. Zephyrus Press, 1975.
Douglas, John Scott, *Caves of Mystery: The Story of Cave Exploration.* Dodd, Mead & Co., 1956.
Ellis, Bryan, *Surveying Caves.* Bridgewater, Somerset: The British Cave Research Association, 1976.
Eyre, Jim, *The Cave Explorers.* The Stalactite Press, 1981.
Fairbridge, Rhodes W., ed., *The Encyclopedia of Geomorphology,* Vol. 3. Dowden, Hutchinson & Ross, 1968.
Fairbridge, Rhodes W., and Joanne Bourgeois, eds., *The Encyclopedia of Sedimentology.* Dowden, Hutchinson & Ross, 1978.
Farr, Martyn, *The Darkness Beckons: The History and Development of Cave Diving.* London: Diadem Books Ltd., 1980.
Ford, T. D., and C. H. D. Cullingford, eds., *The Science of Speleology.* Academic Press, 1976.
Forwood, W. Stump, *An Historical and Descriptive Narrative of the Mammoth Cave of Kentucky.* J. B. Lippincott & Co., 1870.
Franke, Herbert W., *Wilderness under the Earth.* Transl. by Mervyn Savill. London: Lutterworth Press, 1958.
Gurnee, Russell H., *Discovery of Luray Caverns, Virginia.* R. H. Gurnee, Inc., 1978.
Gurnee, Russell and Jeanne, *Gurnee Guide to American Caves.* Zephyrus Press, 1980.
Habe, France, *The Postojna Caves.* Postojna: Postojnska jama, 1981.

Halliday, William R.:
Adventure Is Underground. Harper & Brothers, 1959.
American Caves and Caving. Harper & Row, 1974.
Depths of the Earth: Caves and Cavers of the United States. Harper & Row, 1976.
Hanbury-Tenison, Robin, *Mulu: The Rain Forest.* London: Weidenfeld and Nicolson, 1980.
Herak, M., and V. T. Stringfield, eds., *Karst: Important Karst Regions of the Northern Hemisphere.* Elsevier Publishing Co., 1972.
Hill, Carol A., *Cave Minerals.* National Speleological Society, 1976.
Hogg, Garry, *Deep Down: Great Achievements in Cave Exploration.* Criterion Books, 1962.
Hovey, Horace Carter:
Celebrated American Caverns. Johnson Reprint Corporation, 1970.
Guide Book to The Mammoth Cave of Kentucky. Robert Clarke & Co., 1891.
Jakucs, László, *Morphogenetics of Karst Regions.* Transl. by B. Balkay. Bristol: Adam Hilger Ltd., 1977.
Jasinski, Marc, *La spéléologie.* Paris: Dargaud Éditeur, 1966.
Jennings, J. N., *Karst.* The M.I.T. Press, 1971.
Johnson, Peter, *The History of Mendip Caving.* Newton Abbot, Devon: David & Charles, 1967.
Judson, David, and Arthur Champion, *Caving and Potholing.* Granada Publishing Ltd., 1981.
Lawrence, Joe, Jr., and Roger W. Brucker, *The Caves Beyond.* Zephyrus Press, 1975.
Leroi-Gourhan, André, *Treasures of Prehistoric Art.* Harry N. Abrams, no date.
Lobeck, A. K., *Geomorphology: An Introduction to the Study of Landscapes.* McGraw-Hill, 1939.
Long, Abijah and Joe N., *The Big Cave.* Cushman Publications, 1956.
Lovelock, James, *Caving.* London: B. T. Batsford Ltd., 1969.
McClurg, David R., *Exploring Caves: A Guide to the Underground Wilderness.* Stackpole Books, 1980.
Martel, E.-A.:
Les Abîmes: Les Eaux Souterraines, Les Cavernes, Les Sources. Paris: Librairie Charles Delagrave, 1894.
L'Aven Armand: Description Géologie Historique. Millau: Editions Artières, 1962.
Meloy, Harold, *Mummies of Mammoth Cave.* Micron Publishing Co., 1977.
Minvielle, Pierre, *Grottes et Canyons: Les 100 Plus Belles Courses et Randonnées.* Paris: Éditions Denoël, 1977.
Mohr, Charles E., *The World of the Bat.* J. B. Lippincott, 1976.
Mohr, Charles E., and Thomas L. Poulson, *The Life of the Cave.* McGraw-Hill, 1966.
Mohr, Charles E., and Howard N. Sloane, eds., *Celebrated American Caves.* Rutgers University Press, 1955.
Moore, George W., and Brother G. Nicholas, *Speleology: The Study of Caves.* D. C. Heath and Co., 1964.
Murray, Robert K., and Roger W. Brucker, *Trapped!* G. P. Putnam's Sons, 1979.
Nicod, Jean, *Pays et Paysages du Calcaire.* Paris: Presses Universitaires de France, 1972.
Notice sur les Travaux Scientifiques de M. Édouard-Alfred Martel. Paris: Libraires de l'Académie de Médecine, 1911.
Owen, Luella Agnes, *Cave Regions of the Ozarks and Black Hills.* Johnson Reprint Corporation, 1970.
Palmer, Arthur N., *A Geological Guide to Mammoth Cave National Park.* Zephyrus Press, 1981.
Partington, J. R., *A Short History of Chemistry.* Harper & Brothers, 1960.
Šajn, Srečko, ed., *Postojnska Jama.* Transl. by Zdenka Šlenc. Postojna: Postojnska jama THO, 1978.
Scheffel, Richard L., and Susan J. Wernert, eds., *Natural Wonders of the World.* The Reader's Digest Association, 1980.
Shaw, Trevor R., *History of Cave Science.* Crymych, Wales: Anne Oldham, 1979.
Sieveking, Ann and Gale, *The Caves of France and Northern Spain: A Guide.* London: Vista Books, 1962.
Siffre, Michel:
Dans les Abîmes de la Terre. Paris: Flammarion, 1975.
Grottes, Gouffres & Abîmes. Paris: Hachette Réalités, 1981.
Siffre, Michel, and Georges Dupont, *Les Animaux de Gouffres et des Cavernes.* Paris: Hachette, 1979.
Sloane, Bruce, ed., *Cavers, Caves and Caving.* Rutgers University Press, 1977.
Sparks, B. W., *Geomorphology.* London: Longmans, Green and Co., Ltd., 1960.
Stenuit, Robert, and Marc Jasinski, *Caves and the Marvellous World beneath Us.* A. S. Barnes and Co., 1966.
Sweeting, Marjorie M., *Karst Landforms.* Columbia University Press, 1973.
Taylor, Bayard, *At Home and Abroad.* G. P. Putnam, 1862.
Thompson, Ralph Seymour, *The Sucker's Visit to the Mammoth Cave.* Johnson Reprint Corporation, 1970.
Verne, Jules, *Journey to the Center of the Earth.* Dodd, Mead & Co., 1979.
Waltham, A. C., *The World of Caves.* London: Orbis Publishing Ltd., 1976.
Waltham, Tony, *Caves.* Crown Publishers, 1977.
Weaver, Dwight H., *Onondaga: The Mammoth Cave of Missouri.* Discovery Enterprises, 1973.
Yalden, D. W., and P. A. Morris, *The Lives of Bats.* Quadrangle/The New York Times Book Co., 1975.

Periodicals

The British Caver. Vols. 67-84, Christmas 1977-Spring 1982.
Bulletin of the National Speleological Society. June 1940-December 1981.
Burman, Ben Lucien, "Kentucky's Crazy Cave." *Collier's,* June 6, 1953.
Casteret, Norbert, "Discovering the Oldest Statues in the World." *National Geographic,* August 1924.
"Caves and Caving." *The Bulletin of the British Cave Research Association,* Nos. 1-13, August 1978-August 1981.
Caving International Magazine. Nos. 1-12, October 1978-July 1981.
Crowther, Patricia:
"Discovering the World's Biggest Cave." *The Saturday Review,* April 1973.
"Into Mammoth Cave—The Hard Way." *National Parks & Conservation Magazine,* January 1973.
Detjen, Jim, and Jim Adams, "Hidden River Cave Was Unable to Hide from Man's Pollution." *The Courier-Journal* (Louisville), December 2, 1979.
Durand, J. P., and A. Vandel, "Proteus: An Evolutionary Relic." *Science Journal,* February 1968.
GEO² (Publication of Cave Geology and Geography Section at the National Speleological Society), 1974-1981.
Gilbert, Bil, "Batty about Caves." *Sports Illustrated,* March 15, 1982.
Griffin, Donald R., "Mystery Mammals of the Twilight." *National Geographic,* July 1946.
Hapgood, Fred, "The Ghostly Wings of Night." *GEO,* July 1981.
Herald, Earl S., "Texas Blind Salamander in the Aquarium." *Aquarium Journal,* August 1952.
The Journal of Spelean History, 1968-1981.
Kurtén, Björn, "The Cave Bear." *Scientific American,* March 1972.
Lee, Willis T.:
"New Discoveries in Carlsbad Cavern." *National Geographic,* September 1925.
"A Visit to Carlsbad Cavern." *National Geographic,* January 1924.
Lesy, Michael, "Dark Carnival: The Death of Floyd Collins." *American Heritage,* October 1976.
Marshack, Alexander, "Exploring the Mind of Ice Age Man." *National Geographic,* January 1975.
Martel, E.-A.:
"British Caves and Speleology." *The Geographic Journal,* July-December 1897.
"The Descent of Gaping Ghyll (Yorkshire): A Story of Mountaineering Reversed." *The Alpine Journal,* May 1896.
"Into the Earth's Depths: Twenty Years of Cave-Exploring." *Sunday Magazine,* January 28, 1906.
"The Land of the Causses: The Canon of the Tarn, Montpellier-le-Vieux." *Appalachia: The Journal of the Appalachian Mountain Club,* 1893-1895.
"Speleology: A Modern Sporting Science," *The Yorkshire Ramblers' Club Journal,* 1903-1908.
"Speleology, or Cave Exploration." *Appleton's Popular Science Monthly,* December 1898.
Mohr, Charles E.:
"Exploring America Underground." *National Geographic,* June 1964.
"I Explore Caves." *Natural History,* April 1939.
"Ozark Cave Life." *National Speleological Society Bulletin,* November 1950.
Nicholas (Sullivan), Brother G., "Entrance, Twilight and Dark." *Natural History,* April 1971.
NSS News, January 1941-April 1982.
Oster, Gerald, "The Modern Look of Ice Age Art." *Natural History,* October 1978.
Poulson, Thomas L., "Cave Adaptation in Amblyopsid Fishes." *The American Midland Naturalist,* 1963.
Poulson, Thomas L., and William B. White, "The Cave Environment." *Science,* September 5, 1969.
The Quarterly Journal of the Geological Society of London, 1965.
Ross, Edward S., "Birds That 'See' in the Dark with Their Ears." *National Geographic,* February 1965.
Sainte-Croix, L. de, "E. A. Martel: Explorateur des Abîmes et des Eaux Souterraines." *Revue de l'Alliance Française,* April 1933.
Schreiber, Richard, "The Disaster at Howard's Waterfall Cave." *Georgia Underground,* March-April 1966.
Shields, Mitchell J., "The Lure of the Abyss." *GEO,* June 1979.
Speleo Digest, 1974 and 1978.
Woods, Loren P., "Blind Fishes Found in Cave Pools and Streams." *National Speleological Society Bulletin,* December 1956.
Yager, Jill, "Remipedia, A New Class of Crustacea from a Marine Cave in the Bahamas." *Journal of Crustacean Biology,* 1981.

Other Publications

Barnett, John, "Carlsbad Caverns National Park, New Mexico." Carlsbad Caverns Natural History Association pamphlet, 1979.
Beck, Barry F., ed., *Proceedings of the Eighth International Congress of Speleology,* Vols. 1 and 2. Meet-

ing of the International Union of Speleology, Bowling Green, Kentucky, July 18-24, 1981.
Davies, W. E., and I. M. Morgan, "Geology of Caves." U.S. Department of the Interior Geological Survey pamphlet.
"Earth Resistivity Exploration." Southwest Research Institute pamphlet, San Antonio, Texas.
Eavis, A. J., ed., *Caves of Mulu '80: The Limestone Caves of the Gunong Mulu National Park, Sarawak.* London: The Royal Geographical Society, 1981.
Monroe, Watson H., "A Glossary of Karst Terminology." Geological Survey Water-Supply Paper 1899-K. United States Printing Office, 1970.
Parzefall, Jakob, Jaques Durand and Bernard Richard, "Aggressive Behavior of the European Cave Salamander *Proteus anguinus.*" Paper presented at the Eighth International Congress of Speleology, Bowling Green, Kentucky, 1981.
Peters, Wendell R., and Richard G. Burdick, "Use of an Automatic Earth Resistivity System for Detection of Abandoned Mine Workings." Society of Mining Engineers of the AIME Preprint No. 81-89. Paper presented at the AIME Annual Meeting, Chicago, February 22-26, 1981.
Webb, Barry, and Antony C. Waltham, unpublished manuscript on Mulu National Park.
White, Jim, "Carlsbad Caverns National Park, New Mexico: Its Early Explorations as Told by Jim White." Privately published pamphlet.
White, William, ed., "Geology and Biology of Pennsylvania Caves." General Geology Report 66, Pennsylvania Geological Survey, Fourth Series, Harrisburg, 1976.
Wilford, G. E., *The Geology of Sarawak and Sabah Caves.* Geological Survey, Borneo Region, Malaysia, Bulletin 6, 1964.

PICTURE CREDITS

The sources for the illustrations that appear in this book are listed below. Credits from left to right are separated by semicolons, from top to bottom by dashes.

Cover: © Michael K. Nichols from Woodfin Camp & Associates. 6, 7: Philippe Crochet, Blois, France. 8: Chip Clark. 9: Francis Le Guen, Paris. 10, 11: Ron Simmons—© 1978 Michael K. Nichols from Woodfin Camp & Associates. 12, 13: © Jay Lurie from Bruce Coleman, Inc.; © C. B. Frith from Bruce Coleman, Inc. 14, 15: Dr. Arthur N. Palmer. 16: Courtesy Austrian National Library, Vienna. 18: Map by Bill Hezlep and Jaime Quintero. 20, 21: Art by Bob Wood. 22: Colin Boothroyd, courtesy Andy Eavis, North Humberside, England. 23: Art by I'Ann Blanchette. 24, 25: Colin Boothroyd, courtesy Andy Eavis, North Humberside, England. 26: Photo Bulloz, courtesy Musée du Petit Palais, Paris. 27: Michael Holford, courtesy Victoria and Albert Museum, London. 29: Jean Vertut, Issy-les-Moulineaux, France. 30, 31: Jean Vertut, Issy-les-Moulineaux, France (2)—Des and Jen Bartlett from Bruce Coleman, Inc.; Ernst H. Kastning; Jean Vertut, Issy-les-Moulineaux, France. 32, 33: Courtesy Austrian National Library, Vienna. 34: By permission of the Houghton Library, Harvard University. 36-40: Library of Congress. 42-49: Joseph Natanson, courtesy Narodnega Muzeja, Ljubljana, Yugoslavia. 50: Scott Lamb. 52: Boyer-Viollet, Paris. 53: Library of Congress. 55: David E. Bunnell. 56, 57: © 1981 Tom Bean—Grant Heilman Photography; Ross Ellis, Sydney. 60: Mary Evans Picture Library, London. 61: Art by I'Ann Blanchette. 62, 63: Library of Congress. 64, 65: Map by Bill Hezlep. 66: David E. Bunnell. 67: Art by I'Ann Blanchette. 68-70: Art by Jaime Quintero. 72: Ernst H. Kastning. 73: Y. and F. M. Callot, Vic-en-Bigorre, France. 75: © 1981 Jim Tuten from Black Star. 76, 77: Yoram Lehmann from Robert Harding Picture Library, London. 78: © 1977 Kenrick L. Day. 79: National Park Service. 80, 81: Elaine Garifine (2); Jacques Choppy, Paris. 82: Dr. Arthur N. Palmer—© 1977 Kenrick L. Day. 83: Philippe Crochet, Blois, France. 84: Luray Caverns, Virginia—Chip Clark—Shostal Associates. 85: Ronal C. Kerbo. 86: Collection LA VIE DU RAIL, © P. L. M. 1982, Paris. 88, 89: National Park Service. 90, 91: Art by Lloyd K. Townsend. 93: © 1973 Russ Kinne from Photo Researchers Inc.—The Cavern Supply Company, Inc. 94, 95: The Cavern Supply Company, Inc., except top right, Ed Cooper. 96: Courtesy Gordon L. Smith. 97: National Park Service/Mammoth Cave National Park. 98: Courtesy Gordon L. Smith. 100: UPI. 101: Art by I'Ann Blanchette. 102: Wide World, courtesy *The New York Times.* 103: Reprinted with permission from *The Courier-Journal* and *The Louisville Times.* 104, 105: Courtesy OEFVW, Vienna. 106: Art by I'Ann Blanchette. 107: © *Illustrated London News,* London. 108: Daniel Roucheux, Montrouge, France. 109: Y. and F. M. Callot, Vic-en-Bigorre, France. 110-117: Art by Jack Unruh. 118: Chip Clark. 120: Art by Frederic F. Bigio from B-C Graphics. 123: Charles E. Mohr. 124, 125: Charles E. Mohr, except far right, Scott Lamb. 126: J. A. L. Cooke from Oxford Scientific Films, Oxford, England. 128: ANIMALS ANIMALS/© Raymond A. Mendez. 130: Scott Lamb. 131: ANIMALS ANIMALS/© Robert W. Mitchell. 132: Tercafs—Jacana, Paris; Library of Congress. 133: Charles E. Mohr—© Marvin E. Newman. 134: © Brian Brake from Photo Researchers Inc. 135: Dennis W. Williams. 136, 137: Chip Clark. 138, 139: Charles E. Mohr; Arlan R. Wiker. 140: Charles E. Mohr. 141: ANIMALS ANIMALS/© Robert W. Mitchell. 142: Photo Le Parisien Libéré, Saint-Ouen, France. 144: Jacques Ertaud, courtesy Éditions Arthaud, Paris. 147: Art by I'Ann Blanchette. 148: Photo Le Parisien Libéré, Saint-Ouen, France. 149: *Paris-Match*/Saulnier, Paris. 151: © Martyn Farr from All Sport/Photographic Ltd., Morden, Surrey, England—Martyn Farr, Powys, Wales. 152: Reprinted with permission from *The Courier-Journal* and *The Louisville Times.* 153: Art by Jaime Quintero. 154, 155: David DesMarais/CRF. 157: Map by Bill Hezlep and John Britt. 158: Southwest Research Institute, San Antonio, Texas—art by Frederic F. Bigio from B-C Graphics. 160, 161: Ernst H. Kastning. 162, 163: John Middleton, Sheffield, England, inset, art by Jaime Quintero. 164: Jean-Pierre Charre, Grenoble, France. 165: Art by Jaime Quintero. 166, 167: Anthony Waltham, Nottingham, England, inset, art by Jaime Quintero. 168, 169: © 1979 George Holton from Photo Researchers Inc., inset, art by Jaime Quintero.

INDEX

Numerals in italics indicate an illustration of the subject mentioned.

B

C

D

E

M

N

O

P

Q

R

S